MINITAB® MANUAL

MARIO F. TRIOLA
Dutchess Community College

to accompany

THE TRIOLA STATISTICS SERIES:

Elementary Statistics, Tenth Edition

Elementary Statistics Using Excel, Third Edition

Essentials of Statistics, Third Edition

*Elementary Statistics Using
the Graphing Calculator*, Second Edition

Mario F. Triola
Dutchess Community College

PEARSON
Addison
Wesley

Boston San Francisco New York
London Toronto Sydney Tokyo Singapore Madrid
Mexico City Munich Paris Cape Town Hong Kong Montreal

Preface

This *Minitab Manual,* 10th edition, is a supplement to textbooks in the Triola Statistics Series (listed below) and *Biostatistics for the Biological and Health Sciences* by Triola/Triola.

- *Elementary Statistics*, 10th edition
- *Essentials of Statistics*, 3rd edition
- *Elementary Statistics Using Excel*, 3rd edition
- *Elementary Statistics Using the Graphing Calculator,* 2nd edition

This manual is based on Minitab Release 14, but references are made to earlier releases when appropriate. This manual can also be used with other versions of Minitab, including the *Student Edition of Minitab, R 14.*

    The associated textbooks are packaged with a CD-ROM that includes Minitab worksheets. The data sets in Appendix B of the textbook are included as Minitab worksheets on that CD-ROM. These Minitab worksheets are also available on the web site http://www.aw-bc.com/triola.

Here are major objectives of this manual/workbook and the Minitab software:

- Describe how Minitab can be used for the methods of statistics presented in the textbook. Specific and detailed procedures for using Minitab are included, along with examples of Minitab screen displays.

- Involve students with a spreadsheet type of software application, which will help them work with other spreadsheet programs they will encounter in their professional work.

- Incorporate an important component of computer usage without using valuable class time required for concepts of statistics.

- Replace tedious calculations or manual construction of graphs with computer results.

- Apply alternative methods, such as simulations, that are made possible with computer usage.

- Include topics, such as analysis of variance and multiple regression, that require calculations so complex that they realistically cannot be done without computer software.

It should be emphasized that this manual/workbook is designed to be a supplement to textbooks in the Triola Statistics Series; it is not designed to be a self-contained statistics textbook. It is assumed throughout this manual/workbook that the theory, assumptions, and procedures of statistics are described in the textbook that is used.

Chapter 1 of this supplement describes some of the important basics for using Minitab. Chapters 2 through 14 in this manual/workbook correspond to Chapters 2 through 14 in *Elementary Statistics*, 10th edition. However, the individual chapter *sections* in this manual/workbook generally do *not* match the sections in the textbook. Each chapter includes a description of the Minitab procedures relevant to the corresponding chapter in the textbook. This cross-referencing makes it very easy to use this supplement with the textbook.

Chapters include illustrations of Minitab procedures as well as detailed steps describing the use of those procedures. It would be helpful to follow the steps shown in these sections so the basic procedures will become familiar. You can compare your own computer display to the display given in this supplement and then verify that your procedure works correctly. You can then proceed to conduct the experiments that follow.

For updates to *Elementary Statistics*, this supplement, other supplements, or Minitab, see the following Web sites:

Web site for *Elementary Statistics*: **http://www.aw-bc.com/triola**
Web site for Minitab: **http://www.minitab.com**

We welcome any comments or suggestions for improving this Minitab manual/workbook. Please send them to the Addison–Wesley Statistics Editor.

The author and publisher are very grateful to Minitab, Inc. for the continuing support and cooperation.

M.F.T.
LaGrange, New York
January, 2006

Contents

1
Basics
of Minitab

1-1 Starting Minitab

Minitab Requirements Minitab Release 14 is the latest version available as of this writing. It can be purchased and downloaded online. As this manual was being written, students could purchase and download Minitab for $99.99, and they could rent Minitab for as little as $25.99 for a semester. Minitab Release 14 requires Windows 98, Windows NT 4, Windows 2000, Windows XP Home, or Windows XP Professional with a 300 MHz processor, 64 MB of RAM, a CD ROM, and 85 MB of hard disk space for full installation. Earlier releases of Minitab can be used with Windows 3.1, Macintosh/Power Macintosh, and DOS microcomputers. In addition to Minitab Release 14 and earlier releases, the *Student Edition of Minitab R 14* can also be used. Although this manual/workbook is based on Minitab Release 14, it can be used for earlier releases of Minitab as well as the *Student Edition of Minitab*.

Starting Minitab: With Minitab installed, you can start it by clicking on the Minitab icon on the startup screen.

You can also start Minitab as follows:

1. Click on the taskbar item of **Start**.

2. Select **Programs**.

3. Select **Minitab 14**.

4. Select **Minitab**.

Session/Data Windows After opening Minitab, you should see a display of its two main windows: the Session window and the Data window (as shown on the top of the following page). The Session window displays results and it also allows you to enter commands. The top of the display on the following page consists of the Session window. The Data window, shown on the bottom of the display on the following page, consists of a spreadsheet for entering data in columns. These spreadsheets are called *worksheets* in Minitab. Minitab allows you to have multiple worksheets in different Data windows.

Session and Data Windows

```
MINITAB - Untitled
File  Edit  Data  Calc  Stat  Graph  Editor  Tools  Window  Help
```

Session

```
——————— 1/1/2006 7:36:58 AM ———————

Welcome to Minitab, press F1 for help.
```

Worksheet 1 *

↓	C1	C2	C3	C4	C5	C6	C7	C8	C9	C10	C11	C12	C13	C14
1														
2														
3														
4														
5														

Installing Data Sets from the CD-ROM   The CD-ROM included with the textbook includes Minitab worksheets containing data sets in Appendix B from the textbook. It might be helpful to copy those worksheets to the Data folder that is created when Minitab is installed. Although this is not necessary, it could make it easier to access those data sets when they are needed. The Minitab worksheets are in the Data Sets folder on the CD-ROM and, when copying to the Minitab Data folder, answer "yes" when asked if it is OK to overwrite the worksheet BEARS.MTW.

1-2 Entering Data

In Minitab, there are three types of data: (1) *numeric* data (numbers), (2) *text* data (characters), and (3) *date/time* data. We will work mostly with numeric data. Also, data can be in the form of a column, a stored constant, or a matrix. We will work with columns of data. Refer to the next page, which shows a portion of a typical Minitab screen. We have entered data values in the first six columns identified as C1, C2, C3, C4, C5, and C6, and we have also entered the names of Red, Orange, Yellow, Brown, Blue, and Green for those six columns. The values are weights (grams) of M&M plain candies, and the columns correspond to the different M&M colors.

Entry of sample data is very easy. Click on the first cell on the desired column, type a number, then press the **Enter** key. Then type a second number and press the **Enter** key again. Continue this process until all of the sample data have been entered. If you make a mistake, simply click on the wrong entry and type the correct value and press **Enter**.

Naming columns: It is wise to enter a name for the different columns of data, so that it becomes easy to keep track of data sets with meaningful names. See the names of Red, Orange, Yellow, Brown, Blue, and Green shown in the Data window below.

Worksheets: It often happens that several columns of data are related by being part of one big and happy data set. The following display shows part of a worksheet that contains the weights of a sample of M&M candies of different colors. A worksheet can be handled as a single entity that contains one or more columns. A worksheet can be named so that it can be saved and retrieved. See Section 1-3 which follows.

Worksheet with Weights of M&Ms

↓	C1	C2	C3	C4	C5	C6
	Red	Orange	Yellow	Brown	Blue	Green
1	0.751	0.735	0.883	0.696	0.881	0.925
2	0.841	0.895	0.769	0.876	0.863	0.914
3	0.856	0.865	0.859	0.855	0.775	0.881
4	0.799	0.864	0.784	0.806	0.854	0.865
5	0.966	0.852	0.824	0.840	0.810	0.865

1-3 Saving Data

After entering an individual column of data or several columns of data, you can *save* the values as a Minitab worksheet as follows.

Minitab Procedure for Saving a Worksheet

1. Click on the main menu item of **File**.

2. Click on **Save Current Worksheet As ...**.

3. You will see a dialog box, such as the one shown below. In the "Save in" box, enter the location where you want the worksheet saved. In the "File name" box, enter the name of the file. If you omit the extension of .MTW, it will be automatically provided by Minitab.

4. Click on **Save**.

Saving a Worksheet

1-4   Retrieving Data from the CD-ROM

The CD-ROM packaged with the textbook has worksheets for the data sets in Appendix B of the textbook. See Appendix B in the textbook for the worksheet names. Here is the procedure for retrieving a worksheet from the CD-ROM:

1.   Click on Minitab's main menu item of **File**.

2.   Click on the subdirectory item of **Open Worksheet**.

3.   You should now see a window like the one shown on the following page.

4.   In the "Look in" box at the top, select the location of the stored worksheets. For example, if the CD-ROM is in drive D and you want to open the Minitab worksheet COLA.MTW, do this:

- In the "Look in" box, select drive **D** (or whatever drive contains the CD-ROM).

- Double click on the folder **Data Sets**.

- Double click on the folder **Minitab**.

- Click on **COLA.MTW** (or any other worksheet that you want).

5. Click on the **Open** bar.

6. You will get a message that a copy of the contents of the file will be added to the current project. Click **OK**. The columns of data should now appear in the Minitab display, and they are now available for use.

Opening a Worksheet

1-5 Printing

We have noted that Minitab involves a screen with two separate parts: The Session window at the top and the Data window at the bottom. When obtaining statistics, graphs, and other results, they are displayed in the Session window at the top. Such results can be printed as follows.

Printing the Session Window or Data Window

1. *Session window:* If you want to print the session window portion of the screen, click on the session window.

    Data window: If you want to print the data portion of the screen, click on any cell in the Data window.

2. Click on the main menu item of **File**.

3. Click on **Print Session Window** or **Print Worksheet**. (You will see only one of these two options. The option that you see is the result of the choice you made in Step 1. If you see "Print Worksheet" but you want to print the Session window, go back to Step 1 and click on the Session window portion of the screen.)

4. If you select Print Worksheet in Step 3, you will get a dialog box. Enter your preferences, then click **OK**.

Printing a Large Data Set in a Word Processor

Most of the data sets used with Triola statistics textbooks are not so large that they cannot be printed on a few pages. If you do enter or somehow create a data set with thousands of values, printing would require many pages. One way to circumvent that problem is to move the data to a word processor where the values can be reconfigured for easier printing. Follow these steps.

1. With the data set displayed in the data window, click on the value at the top. Hold the mouse button down and drag the mouse to the bottom of the data set, then release it so that the entire list of values is highlighted.

2. Click on **Edit**, then click on **Copy Cells**.

3. Now go into your word processor and click on **Edit**, then **Paste**. The entire list of values will be in your word processing document where you can configure them as you please. For example, you might press **End**, then press the **space bar**, then press the **Del** key. The second value will be moved up to the top row. Repeat this process to rearrange the data in multiple rows and columns (instead of one really large column). For example, this approach was used to print on *one* page the 175 values in the column labeled CANS109 that is in the Minitab worksheet named CANS. The result is shown below. You can change the number of columns as you desire.

270 273 258 204 254 228 282 278 201 264 265 223 274 230 250 275 281 271 263 277

275 278 260 262 273 274 286 236 290 286 278 283 262 277 295 274 272 265 275 263

251 289 242 284 241 276 200 278 283 269 282 267 282 272 277 261 257 278 295 270

268 286 262 272 268 283 256 206 277 252 265 263 281 268 280 289 283 263 273 209

259 287 269 277 234 282 276 272 257 267 204 270 285 273 269 284 276 286 273 289

263 270 279 206 270 270 268 218 251 252 284 278 277 208 271 208 280 269 270 294

292 289 290 215 284 283 279 275 223 220 281 268 272 268 279 217 259 291 291 281

230 276 225 282 276 289 288 268 242 283 277 285 293 248 278 285 292 282 287 277

266 268 273 270 256 297 280 256 262 268 262 293 290 274 292

1-6 Command Editor and Transforming Data

Data may be *transformed* with operations such as adding a constant, multiplying by a constant, or using functions such as logarithm (common or natural), sine, exponential, or absolute value. For example, if you have a data set consisting of temperatures on the Fahrenheit scale (such as the body temperatures in Appendix B of the textbook) and you want to transform the values to the Celsius scale, you can use the equation

$$C = \frac{5}{9}(F - 32)$$

Such transformations can be accomplished by using Minitab's *command editor* or by using Minitab's virtual calculator.

Using Command Editor to Transform Data

1. Click on the Session window portion of the screen. (The Session window is on the top portion; the Data window is the bottom portion.)

2. Click on **Editor**.

3. Click on **Enable Commands**.

4. You should now see **MTB >** and you can enter a command. The command
   LET C2 = (5/9)*(C1 — 32)

   tells Minitab to take the values in column C1, subtract 32 from each value, then multiply by the fraction 5/9. This expression corresponds to the above formula for converting Fahrenheit temperatures to the Celsius scale. The temperatures on the Fahrenheit scale of 98.6, 98.6, 98.0, 97.3, and 97.2 are converted to these temperatures on the Celsius scale: 37.0000, 37.0000, 36.6667, 36.2778, and 36.2222.

Using Calc/Calculator to Transform Data

You can also click on **Calc**, then **Calculator** to get a virtual calculator that allows you to create new columns by performing operations with existing columns. The display below shows that column C7 will be the result of adding column C1 to five times the square of each value in column C2. (Multiplication is represented by * and exponents are represented by **. Other functions are available in the "Functions" window of the calculator.)

1-7 Closing Worksheets and Exiting Minitab

Closing worksheets: You can close the current *worksheet* by clicking on the × located at the upper right corner of the worksheet window. If you click on that ×, you will be told that the data in the worksheet will be removed and this action cannot be undone. You will be given the opportunity to save the data in a separate file, or you can click **No** to close the window. The Minitab program will continue to be active, even though you have just closed the current worksheet.

Exiting Minitab: Had enough for now? To exit or quit the Minitab program, click on the × located in the extreme upper right corner. Another way to exit Minitab is to click on **File**, then click on **Exit**. In both cases, you will be asked if you want to save changes to the project before closing. This is your last chance to save items that you want saved. Assuming that you are finished and you want to exit Minitab, click **No**.

1-8 Exchanging Data with Other Applications

There may be times when you want to move data from Minitab to another application (such as Excel or STATDISK or Word) or to move data from another application to Minitab. Instead of manually retyping all of the data values, you can usually transfer the data set directly. Given below are two ways to accomplish this. The first of the following two procedures is easier, if it works (and it often does work).

Method 1: Use Edit/Copy and Edit/Paste

1. With the data set displayed in the source program, use the mouse to highlight the values that you want to copy. (Click on the first value, then hold the mouse button down and drag it to highlight all of the values, then release the mouse button.)

2. While still in the source program, click on **Edit**, then click on **Copy**.

3. Now go into the software program that is your destination and click on **Edit**, then **Paste**. The entire list of values should reappear.

Method 2: Use Text Files

1. In the software program containing the original set of data, create a text file of the data. (In Minitab you can create a text file by clicking on **File**, then **Other Files**, then **Export Special Text**.)

2. Minitab and most other major applications allow you to import the text file that was created. (To import a text file into Minitab, click on **File**, then **Other Files**, then **Import Special Text**.)

CHAPTER 1 EXPERIMENTS: Basics of Minitab

1-1.   ***Entering Sample Data*** When first experimenting with procedures for using Minitab, it's a good strategy to use a small data set instead of one that is large. If a small data set is lost, you can easily enter it a second time. In this experiment, we will enter a small data set, save it, retrieve it, and print it. The Body Temperatures data set from Appendix B includes these body temperatures, along with others:

98.6   98.6   98.0   98.0   99.0   98.4   98.4   98.4   98.4   98.6

a.   Start Minitab and enter the above sample values. (See the procedure described in Section 1-2 of this manual/workbook.)

b.   Save the worksheet using the worksheet name of TEMP (because the values are body temperatures). See the procedure described in Section 1-3 of this manual/ workbook.

c.   Print the data set by printing the display of the data window. (When first loading Minitab, the data window might not display all 10 values simultaneously, but the data window can be enlarged to display more data. Simply position the mouse on the top border of the data window, then hold the mouse button down to click and drag the top border upward.)

d.   Exit Minitab, then restart it and retrieve the worksheet named TEMP. Save another copy of the same data set using the file name of TEMP2. Print TEMP2 and include the title at the top.

1-2.   ***Transforming Data***   Experiment 1-1 results in saving the worksheet TEMP that contains body temperatures in degrees Fahrenheit. Retrieve that data set, then proceed to transform the temperatures to the Celsius scale. Store the Celsius temperatures in column C2. (See Section 1-6 in this manual/workbook.) Print the worksheet containing the original Fahrenheit temperatures in column C1 and the corresponding Celsius temperatures in column C2.

1-3.   ***Retrieving Data*** The worksheet FHEALTH.MTW (female health) is already stored on the CD-ROM that is included with the textbook. That worksheet contains 13 individual columns of health-related measurements from 40 randomly selected women. Open the worksheet FHEALTH.MTW, then print its contents.

1-4.   ***Transforming Data***   From the Minitab worksheets stored on the CD-ROM that is packaged with the textbook, retrieve the worksheet named BEARS.MTW, which lists measurements from a sample of bears. Column C9 (WEIGHT) consists of the weights (in pounds) of the bears. To convert the weights to kilograms, multiply them by 0.4536. Use Minitab to convert the weights from pounds to kilograms. In the space below, write the weights (in kilograms) of the first five bears. Print the screen display showing at least the first few weights converted to kilograms.

1-5.   ***Generating Random Data***   In addition to entering or retrieving data, Minitab can also *generate* data sets. In this experiment, we will use Minitab to simulate the rolling of a single die 500 times.

     a. Select **Calc** from the main menu bar, then select **Random Data.**

     b. Select **Integer**, because we want whole numbers.

     c. Enter 500 for the number of values, enter the column (such as C1) to be used for the data, enter 1 for minimum, and enter 6 for the maximum, then click **OK.**

     d. Examine the displayed values and count the number of times that 5 occurs. Record the result here: _____

1-6.   ***Generating Random Data***   Experiment 1-5 results in the random generation of the digits 1, 2, 3, 4, 5, and 6. Minitab can also generate sample data from other types of populations. In this experiment, we will use Minitab to simulate the random selection of 50 IQ scores.

     a. Select **Calc** from the main menu bar, then select **Random Data.**

     b. Select **Normal**. (The normal distribution is described in the textbook.)

     c. Enter 50 for the number of values, enter the column (such as C1) to be used for the data, enter 100 for the mean, and enter 15 for the standard deviation. (These statistics are described in the textbook.) Click **OK** and print the results.

1-7.   ***Saving a Worksheet***   Listed below are the ages of actresses and actors at the time they won Oscars. Enter the ages of the winning actresses in column C1 and name that column Actresses. Enter the ages of the winning actors in column C2 and name that column Actors. Save the worksheet with the name OSCAR. Print the worksheet.

Actresses

22	37	28	63	32	26	31	27	27	28	30	26
29	24	38	25	29	41	30	35	35	33	29	38
54	24	25	46	41	28	40	39	29	27	31	38
29	25	35	60	43	35	34	34	27	37	42	41
36	32	41	33	31	74	33	50	38	61	21	41
26	80	42	29	33	35	45	49	39	34	26	25
33	35	35	28								

Actors

44	41	62	52	41	34	34	52	41	37	38	34
32	40	43	56	41	39	49	57	41	38	42	52
51	35	30	39	41	44	49	35	47	31	47	37
57	42	45	42	44	62	43	42	48	49	56	38
60	30	40	42	36	76	39	53	45	36	62	43
51	32	42	54	52	37	38	32	45	60	46	40
36	47	29	43								

2
Graphing Data

Important note: The topics of this chapter require that you use Minitab to enter data, retrieve data, save files, and print results. These functions are described in Chapter 1 of this manual/ workbook. Be sure to understand those functions from Chapter 1 before beginning this chapter.

Important Characteristics of Data

The textbook states that when describing, exploring, and comparing data sets, the following characteristics are usually extremely important:

1. **Center:** Measure of center, which is a representative or average value that gives us an indication of where the middle of the data set is located

2. **Variation:** A measure of the amount that the values vary among themselves

3. **Distribution:** The nature or shape of the distribution of the data, such as bell-shaped, uniform, or skewed

4. **Outliers:** Sample values that are very far away from the vast majority of the other sample values

5. **Time:** Changing characteristics of the data over time

In this chapter, we learn how to use Minitab as a tool for investigating the *distribution* of the data by constructing suitable graphs.

Exercise 1-7 from Chapter 1 in this manual/workbook lists ages of actresses and actors when they won Oscars. If that exercise was successfully completed, those ages are stored as two columns of the worksheet named OSCAR. We will proceed to use that worksheet in showing how Minitab can be used to generate important graphs. We begin with histograms.

2-1 Histograms

The Triola textbook describes histograms and provides detailed procedures for constructing them. It is noted that a histogram is an excellent device for exploring the *distribution* of a data set. Here is the Minitab procedure for generating a histogram.

1. Enter the data in a Minitab column, such as column C1. If the data are already stored in a Minitab worksheet, retrieve the worksheet using the procedure described in Section 1-4 of this manual/workbook.

2. Click on the main menu item of **Graph**.

3. Select **Histogram** from the subdirectory. (Also, if using Minitab Release 14, click on the option of **Simple** and click **OK**.)

4. In the dialog box, click on the column with the data so that the column label (such as C1) or the data name (such as Actresses) appears in the box as shown.

Generating a Histogram

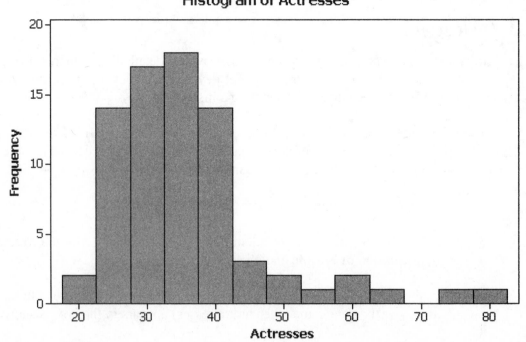

5.   Using Minitab with the ages of the Oscar-winning actresses results in the following histogram. You can edit a histogram or any other graph by double-clicking on the attribute that you want to change. To change the horizontal scale of a histogram, double-click on the *x*-axis and use the pop-up window to change the scale. For example, in the Edit Scale pop-up window, click on the **binning** tab, select **Midpoint**, select **Midpoint/Cutpoint positions**, and enter the desired class midpoint values (such as 20 30 40 50 60 70 80).

Histogram of Actresses

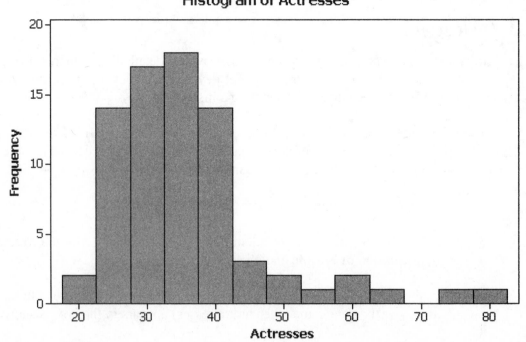

2-2 Dotplots

Here is the Minitab procedure for creating a dotplot.

1. Enter or retrieve the data into a Minitab column.

2. Click on the main menu item of **Graph**.

3. Select **Dotplot**. (Also, if using Minitab Release 14, select the **Simple** dotplot and click **OK**.)

4. The dialog box will list the columns of data at the left. Click on the desired column. You can edit a dotplot or any other graph by double-clicking on the attribute you want to modify. To change the scale, double-click on the axis and use the pop-up window.

5. Click **OK**.

Shown below is the Minitab dotplot display of the ages of actresses listed in Exercise 1-7 of this manual/workbook. The original dotplot showed values of 24, 32, . . . , 80 along the horizontal scale, but the dotplot was modified to show the more convenient values of 20, 30, . . . , 80 instead. This change in the numbers used for the horizontal scale was made by double-clicking on a number on the horizontal scale, selecting the **scale** tab, then manually entering the values of 20, 30, . . . , 80 in the box labeled **Position of ticks**.

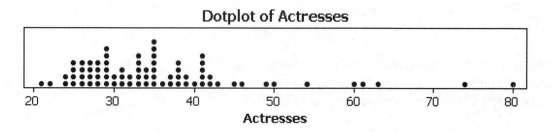

2-3 Stemplots

Minitab can provide stemplots (or stem-and-leaf plots). The textbook notes that a **stemplot** represents data by separating each value into two parts: the stem (such as the leftmost digit) and the leaf (such as the rightmost digit). Here is the Minitab procedure for generating a stemplot.

1. Enter or retrieve the data into a Minitab column.

2. Click on the main menu item of **Graph**.
3. Select **Stem-and-Leaf**.

4. The dialog box lists columns of data at the left. Click on the desired column.

5.   (If using Minitab Release 14, remove the check mark for "Trim outliers," so
        that outliers will be included. Also, enter a desired value for the interval.) Click
        OK.

Shown below is the Minitab stem-and-leaf plot that results from the ages of actresses listed in
Exercise 1-7 of this manual/workbook. The actual stem–and–leaf plot is highlighted here with a
bold font.

```
Stem-and-leaf of Actresses   N  = 76
Leaf Unit = 1.0

    4     2  1244
   26     2  5555666677778888999999
  (15)    3  001112233333444
   35     3  55555555677888899
   19     4  011111223
   10     4  569
    7     5  04
    5     5
    5     6  013
    2     6
    2     7  4
    1     7
    1     8  0
```

The above stemplot is the default that is automatically generated by Minitab. In this case, it is an
expanded stemplot with classes of 20-24, 25-29, 30-34, 35-39, and so on. To obtain a stemplot
that is not expanded, enter an increment value of 10.

Reading and Interpreting a Minitab Stem–and–Leaf Plot: In addition to the stemplot itself,
Minitab also provides another column of data at the extreme left as shown above. In this Minitab
display, *the leftmost column represents cumulative totals*. The left column above shows that
there are 4 sample values included between 20 and 24, and there are 26 values between 20 and
29. The left column entry of (15) indicates that there are 15 data values in the row containing the
median. The left-column entries *below* the median row represent cumulative totals from the
bottom up, so that the 1 at the bottom indicates that there is one value between 80 and 84.
Similarly, the 7 in the left column indicates that there are seven values between 50 and 84.

2-4 Pareto Charts

The textbook describes a Pareto chart as a bar graph for categorical data, with the bars arranged
in order according to frequencies. The tallest bar is at the left, and the smaller bars are farther to

the right. Shown below is a table of categorical data, and the Minitab Pareto chart follows the table. The line graph at the top of Minitab's Pareto chart represents cumulative totals.

Phone Complaint	Number
Rates and Services	4,473
Marketing	1,007
International Calling	766
Access Charges	614
Operator Services	534
Slamming	12,748
Cramming	1,214

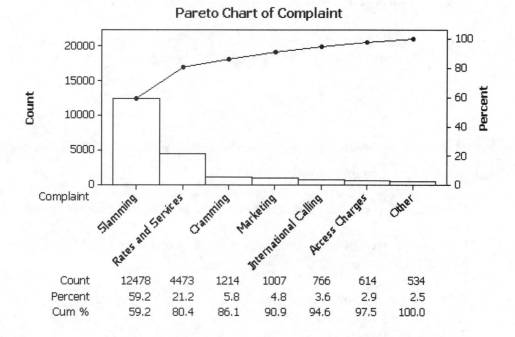

Pareto Chart of Complaint

	Slamming	Rates and Services	Cramming	Marketing	International Calling	Access Charges	Other
Count	12478	4473	1214	1007	766	614	534
Percent	59.2	21.2	5.8	4.8	3.6	2.9	2.5
Cum %	59.2	80.4	86.1	90.9	94.6	97.5	100.0

Here is the procedure for using Minitab to construct a Pareto chart:

1. Enter the labels (such as Rates and Services, Marketing, etc.) in column C1.

2. Enter the corresponding frequency counts in column C2. The worksheet for the above table should appear as shown below.

↓	C1-T	C2
	Complaint	**Frequency**
1	Rates and Services	4473
2	Marketing	1007
3	International Calling	766
4	Access Charges	614
5	Operator Services	534
6	Slamming	12478
7	Cramming	1214

3.   Select the menu items of **Stat**, **Quality Tools**, and **Pareto Chart**.

4.   When the dialog box is displayed, select the **Charts defect table** option and enter C1 in the labels box and C2 in the frequency box.

5.   Click **OK**.

2-5 Pie Charts

To obtain a pie chart from Minitab, follow this procedure.

1.   Enter the labels (such as Rates and Services, Marketing, etc.) in column C1.

2.   Enter the corresponding frequency counts in column C2. (See the preceding Minitab screen display.)

3.   Select the main menu item of **Graph**.

4.   Select **Pie Chart**.

5.   When the dialog box is displayed, select the **Charts defect table** (in Minitab Release 14, select **Chart values from a table**) option and enter C1 for the labels (or categorical variable) box and enter C2 in the frequency (or summary variables) box.

6.   Click **OK**. Shown below is the Minitab pie chart corresponding to the table of phone complaint data on the preceding page.

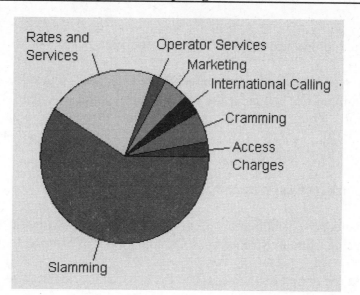

2-6 Scatterplots

The textbook describes a scatterplot (or scatter diagram). A scatterplot can be very helpful in seeing a relationship between two variables. The scatterplot shown below results from paired data consisting of the number of chirps per minute by a cricket and the temperature. This scatterplot shows that as the number of chirps increases, the corresponding temperature tends to be higher.

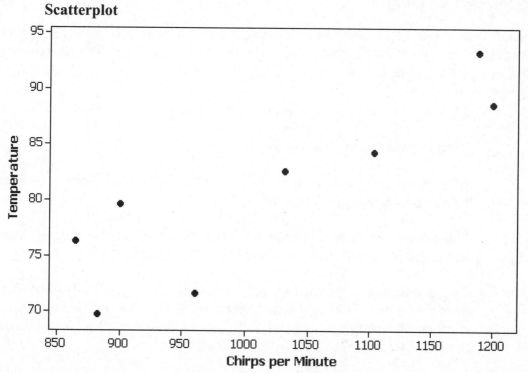

Here is the Minitab procedure for generating a scatterplot.

1. Given a collection of paired data, enter the values for one of the variables in column C1, and enter the corresponding values for the second variable in column C2. (You can also use any two Minitab columns other than C1 and C2.) Be careful to enter the two columns of data so that they are matched in the same way that they are paired in the original listing.

2. Click on **Graph** from the main menu.

3. Select the subdirectory item of **Plot** (or **Scatterplot** in Minitab Release 14). If using Minitab Release 14, select the option of **Simple**, then click **OK**.

4. In the dialog box, enter C1 in the box for the Y variable, then enter C2 in the box for the X variable (or vice versa).

5. Click **OK**.

Another procedure for generating a scatterplot is to select **Stat**, then **Regression**, then **Fitted Line Plot**. We will use this option later when we discuss the topics of correlation and regression. For now, we are simply generating the scatter diagram to visually explore whether there is an obvious pattern that might reveal some relationship between the two variables.

2-7 Time–Series Graph

Time–series data are data that have been collected at different points in time, and Minitab can graph such data so that patterns become easier to recognize. Here is the Minitab procedure for generating a time–series graph.

1. Given a collection of time–series data, enter the sequential values in column C1.

2. Select the main menu item of **Graph**.

3. Select the menu item of **Time Series Plot**. (If using Minitab Release 14, select the option of **Simple** and click **OK**.)

4. A dialog box now appears. Enter C1 in the first cell under Y in the "Graph" list or, if using Minitab Release 14, enter C1 in the "Series" box.

5. Enter a start time to be used for the horizontal axis. Click the **Options** button to enter a start time; if using Minitab Release 14, click **Time/Scale** and proceed to select the desired time scale (such as "Calendar") along with a start value (such as 1980 and increment value (such as 1). Click **OK** twice.

Listed below are the numbers of runway near-hits, listed in order for each year beginning with 1990 (based on data from the Federal Aviation Administration), and the Minitab time-series graph follows. When compared to the list of frequency values, we can see that the time-series graph does a much better job of showing the upward trend that has been occurring in recent years.

281 242 219 186 200 240 275 292 325 321 421

Time-Series Graph of Runway Near Misses

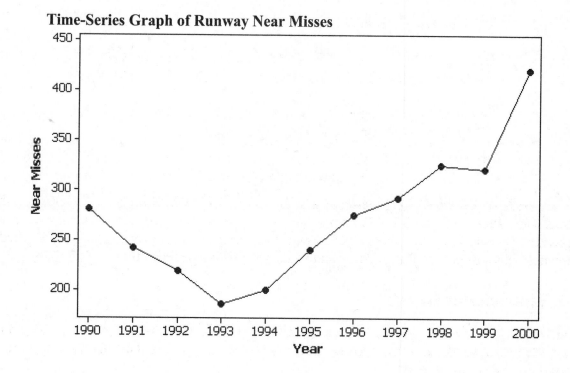

This chapter has described procedures for using Minitab to construct histograms, dotplots, stemplots, Pareto charts, pie charts, scatter diagrams, and time-series graphs. Boxplots are also helpful graphs, and they will be described in Chapter 3 of this manual/workbook. See Section 3-3.

CHAPTER 2 EXPERIMENTS: **Graphing Data**

2–1.   ***Histogram***   Appendix B includes Data Set 1 with measurements from 40 males and 40 females. The Minitab worksheets MHEALTH (male health) and FHEALTH (female health) contain those measurements. Use the 40 heights of males to generate a Minitab histogram and print the result.

2-2.   ***Histogram***   Repeat Experiment 2-1, but instead of using Minitab's default values for the horizontal axis, use the midpoint values of 55, 60, 65, 70, 75. *Hint:* Double-click on a value on the horizontal axis, click on the **binning** tab, select the interval type of **Midpoint**, then select **Midpoint/Cutpoint positions** and enter the midpoint values in the box.

2-3.   ***Histogram***   Repeat Experiment 2-1 using the 40 heights of females.

2-4.   ***Histogram***   Repeat Experiment 2-1 using the 40 heights of females, but instead of using Minitab's default values for the horizontal axis, use the midpoint values of 55, 60, 65, 70, 75. *Hint:* Double-click on a value on the horizontal axis, click on the **binning** tab, select the interval type of **Midpoint**, then select **Midpoint/Cutpoint positions** and enter the midpoint values in the box.

2-5.   ***Comparing Histograms***   Because the histograms generated in Experiments 2-2 and 2-4 use the same scale for the horizontal axis, it is much easier to compare the histograms. Compare the two histograms and write conclusions here:

_____

_____

_____

2-6.   ***Dotplot***   Use the 40 heights of males described in Experiment 2-1 to generate a dotplot. Instead of using Minitab's default scale, use a horizontal scale with the values of 55, 60 65, 70, 75, 80. *Hint:* After generating the dotplot, double-click on a value on the horizontal scale, select the **scale** tab, and proceed to enter 55, 60, 65, 70, 75, 80 for the positions of the ticks. Print the result.

2-7.   ***Dotplot***   Repeat Experiment 2-6 using the 40 heights of the females.

2-8.   ***Comparing Dotplots***   Because the dotplots generated in Experiments 2-6 and 2-7 use the same scale for the horizontal axis, it is much easier to compare them. Compare the two dotplots and write conclusions here:

_____

_____

2-9.   ***Stemplot***   Use the 40 heights of males described in Experiment 2-1 to generate a stemplot. Print the result.

2-10. ***Stemplot*** Use the 40 heights of females described in Experiment 2-1 to generate a stemplot. Print the result.

2-11. ***Comparing Stemplots*** Compare the stemplots from Experiments 2-9 and 2-10 and write conclusions here:

_____

_____

2-12. ***Interpreting Stemplot*** A Minitab-generated stemplot is shown below.

Enter the first five values represented in the stemplot: _____

What does the value of 3 in the left column indicate?

_____

_____

What does the value of 2 in the left column indicated?

_____

_____

What does the value of 5 in the left column indicate?

_____

_____

```
    1     4     9
    1     5
    1     6
    3     7     39
    7     8     1345
   (2)    9     59
    7    10
    7    11     12
    5    12     7
    4    13     0588
```

2-13. ***Scatterplots*** Appendix B includes Data Set 1 for measurements from 40 males and 40 females. The Minitab worksheets MHEALTH (male health) and FHEALTH (female health) contain those measurements. Use the heights and pulse rates of males to generate a Minitab scatterplot and print the result. Based on the result, does there appear to be a relationship between the heights of males and their pulse rates? Explain.

_____

_____

2-14. *Scatterplots* Appendix B includes a data set listing measurements from bears. Use the lengths and chest measurements of the bears to generate a Minitab scatterplot and print the result. Based on the result, does there appear to be a relationship between the lengths of bears and their chest sizes? Explain.

_____

_____

2–15. *Pareto Chart* A study was conducted to determine how people get jobs. The table lists data from 400 randomly selected subjects. The data are based on results from the National Center for Career Strategies. Use Minitab to construct a Pareto chart that corresponds to the given data. Print the result. If someone would like to get a job, what seems to be the most effective approach?

_____

_____

Job Sources of Survey Respondents	Frequency
Help-wanted ads	56
Executive search firms	44
Networking	280
Mass mailing	20

2–16. *Pie Chart* Refer to the data given in Experiment 2–15 and use Minitab to construct a pie chart. Print the result. Compare the pie chart to the Pareto chart. Which graph is more effective in showing the relative importance of job sources?

_____

_____

2–17. ***Time–Series Graphs***   Given below are the numbers of indoor movie theaters, listed in order by row, for each year beginning with 1987 (based on data from the National Association of Theater Owners). Use Minitab to construct a time-series graph, then print the result. What is the trend?

20,595   21,632   21,907   22,904   23,740   24,344   24,789   25,830   26,995
28,905   31,050   33,418   36,448   35,567   34,490   35,170   35,361

_____

_____

2-18. ***Combining Data***   Open the Minitab worksheet M&M.MTW, which consists of six different columns of data listing the weights (grams) of M&M candies arranged according to color. Use **Data/Stack** (or **Manip/Stack** if using an earlier version of Minitab) to make a copy of the data all stacked together in column C7 of the current worksheet. Print a histogram of the data set. Describe the important characteristics of the data set. Be sure to address the nature of the distribution.

_____

_____

2–19. ***Working with Your Own Data***   Through observation or experimentation, collect your own set of sample values. Obtain at least 40 values and try to select data from an interesting population. Use Minitab to generate graphs suitable for describing the distribution of the data. Describe the data and important characteristics.

_____

_____

_____

_____

_____

3

Statistics for Describing, Exploring, and Comparing Data

Important note: The topics of this chapter require that you use Minitab to enter data, retrieve data, save files, and print results. These functions are described in Chapter 1 of this manual/ workbook. Be sure to understand those functions from Chapter 1 before beginning this chapter.

The textbook notes that these characteristics of a data set are extremely important: center, variation, distribution, outliers, and changing characteristics of data over time. Chapter 2 of this manual/workbook presented a variety of graphs that are helpful in learning about the distribution of a data set. This chapter presents methods for learning about center, variation, and outliers.

Exercise 1-7 from Chapter 1 of this manual/workbook includes the ages of actresses and actors at the time they won Oscars. That exercise required that the data be saved as a Minitab worksheet with two columns named Actresses and Actors. The data are reproduced below. We will use this data set to illustrate the methods of this chapter.

Actresses

22	37	28	63	32	26	31	27	27	28	30	26
29	24	38	25	29	41	30	35	35	33	29	38
54	24	25	46	41	28	40	39	29	27	31	38
29	25	35	60	43	35	34	34	27	37	42	41
36	32	41	33	31	74	33	50	38	61	21	41
26	80	42	29	33	35	45	49	39	34	26	25
33	35	35	28								

Actors

44	41	62	52	41	34	34	52	41	37	38	34
32	40	43	56	41	39	49	57	41	38	42	52
51	35	30	39	41	44	49	35	47	31	47	37
57	42	45	42	44	62	43	42	48	49	56	38
60	30	40	42	36	76	39	53	45	36	62	43
51	32	42	54	52	37	38	32	45	60	46	40
36	47	29	43								

3-1 Descriptive Statistics

Given a data set, such as the above list of ages of actresses when they won Oscars, we can use Minitab to obtain descriptive statistics, including the mean, standard deviation, and quartiles. Here is the Minitab procedure.

1.   Enter or retrieve the data in a Minitab column, such as C1. (See Section 1-2.)

2.   Click on the main menu item of **Stat**.

3.   Click on the subdirectory item of **Basic Statistics**.

4.   Select **Display Descriptive Statistics**.

5.   A "Display Descriptive Statistics" dialog box will pop up. In the "Variables" box enter the column containing the data that you are investigating. You can manually enter the column, such as C1 or the name of the column, or you can click on the desired column displayed at the left.

- You can also click on the **Graphs** bar to generate certain graphs, including a histogram and boxplot.
- You can also click on the **Statistics** bar to select the particular statistics that you want included among the results.

6.   Click **OK**.

If you enter the ages of the actresses listed on the preceding page and follow the above procedure, you will get a Minitab display like the one shown below. For this display, the author clicked on the **Statistics** bar and selected the statistics that are particularly relevant to discussions in the textbook.

Descriptive Statistics

Variable	Total Count	Mean	StDev	Variance	CoefVar		
Actresses	76	35.68	11.06	122.41	31.00		

Variable	Minimum	Q1	Median	Q3	Maximum	Range
Actresses	21.00	28.00	33.50	39.75	80.00	59.00

From the above Minitab display, we obtain the following important descriptive statistics (expressed by applying the round-off rule of using one more decimal place than in the original data values, and also including the appropriate units).

- Number of data values: 76
- Mean: $\bar{x}$ = 35.7 years
- Standard deviation: s = 11.1 years
- Variance: s^2 = 122.4 year^2
- Range: 59.0 years
- Coefficient of variation: 31.0
- Five-number-summary:
    Minimum: 21.0 years
    First quartile Q_1: 28.0 years
    Median: 33.5 years
    Third quartile Q_3: 39.8 years
    Maximum: 80.0 years

Quartiles: The textbook uses a simplified procedure for finding quartiles, so the quartiles Q_1 and Q_3 found with Minitab may differ slightly from those found by using the textbook method. For the above ages of actresses, Minitab's first quartile is the same value that would be obtained by using the method described in the textbook, but Minitab's third quartile of 39.8 years is slightly different from the value of 39.5 years obtained by using the procedure in the textbook.

3-2 *z* Scores

The textbook describes *z* scores (or standard scores). For sample data with mean $\bar{x}$ and standard deviation *s*, the *z* score can be found for a sample value *x* by computing

$$z = \frac{x - \bar{x}}{s} \quad \text{or} \quad z = \frac{x - \mu}{\sigma}$$

Here is the Minitab procedure for finding *z* scores corresponding to sample values. (This procedure uses column C1, but any other Minitab column can be used instead.)

1.   Enter a list of data values in column C1.

2.   Click on the main menu item of **Calc**.

3.   Select **Standardize** from the subdirectory.

4.   A dialog box should appear. Enter C1 for the input column and C2 for the column in which the results will be stored.

5.   There are buttons for different calculations, but be sure to select the button for "Subtract mean and divide by standard deviation."

6.   Click **OK**, and column C2 will magically appear with the *z* score equivalent of each of the original sample values. Because column C2 is now available as a set of data, you can explore it using Descriptive Statistics, Histogram, Dotplot, and so on.

Shown below are the first few rows from the Minitab display of the *z* scores for the ages of Actresses. The column label of *z* was manually entered.

Actresses	z
22	-1.23686
37	0.11893
28	-0.69454

3-3 Boxplots

The textbook describes the construction of boxplots based on the minimum value, maximum value, median, and quartiles Q_1 and Q_3. *Note:* The values of Q_1 and Q_3 generated by Minitab may be slightly different from the values obtained by using the procedure described in the textbook. Here is the Minitab procedure for generating boxplots.

1.   Enter or retrieve the data into a column of Minitab.

2.   Click on the main menu item of **Graph**.

3.   Select the subdirectory item of **Boxplot**.

4.   If using Minitab Release 14 or later, select **One Y Simple** for one boxplot or select **Multiple Y's Simple** for boxplots of two or more data sets.

5.   In the dialog box, click on the column(s) containing the data for which the boxplot will be produced.

6.   Click **OK**.

Shown below are the two boxplots for the ages of actresses and actors listed near the beginning of this chapter. Boxplots are particularly useful for comparing data sets. The asterisks are used to identify data values that appear to be outliers. By comparing the two boxplots, we see that the ages of actresses appear to be considerably less than the ages of actors.

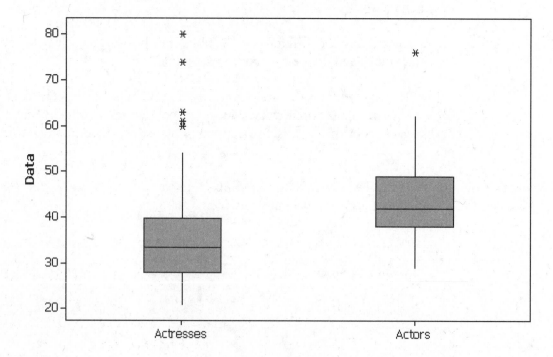

3-4 Sorting Data

There is often a need to *sort* data. Data are sorted when they are arranged in order from low to high (or from high to low). Identification of outliers becomes easier with a sorted list of values. Here is how we can use Minitab to sort data.

1. Enter or retrieve a data set into a column of Minitab.

2. Click on the main menu item of **Data.** [If using a version of Minitab before Release 14, click on the main menu item of **Manip** (for Manipulate)].

3. Click on the subdirectory item of **Sort**.

4. You will now see a dialog box like the one shown below. (The arrangement of items may be somewhat different in earlier versions of Minitab.) In the "Sort column(s)" box, enter the column that you want to sort. In the "By column" box, enter the column to be used as the basis for sorting (this is usually the column containing the original unsorted data). In the "Column(s) of current worksheet" box, enter the column where you want the sorted data to be placed. (You can place the sorted data in the same column as the original data.) The entries shown below tell Minitab to sort the Actresses data (in column C1) and store the sorted list in column C5. Click **OK** after completing the dialog box.

3-5 Outliers

Minitab does not have a specific function for identifying outliers, but some of the functions described in this chapter can be used. One easy procedure for identifying outliers is to *sort* the data as described in Section 3-4 of this manual/workbook, then simply examine the values near the beginning of the list or the end of the list. Consider a value to be an outlier if it is very far away from almost all of the other data values.

　　　Another way to identify outliers is to construct a boxplot. We have noted that Minitab uses asterisks to depict outliers that are identified with a specific criterion that uses the *interquartile range* (IQR). The interquartile range IQR is as follows:

$$IQR = Q_3 - Q_1$$

Minitab identifies data values as outliers and denotes them with asterisks in a boxplot by applying these criteria:

A data value is an outlier if it is …

　　　　　　　　above Q_3 by an amount greater than 1.5 × IQR
　　or　　　**below Q_1 by an amount greater than 1.5 × IQR.**

Remember that important characteristics of data sets are center, variation, distribution, outliers, and time (that is, changing characteristics of data over time). Outliers are important for a variety of reasons. Some outliers are errors that should be corrected. Some outliers are correct values that may have a very dramatic effect on results and conclusions, so it is important to know when outliers are present so that their effects can be considered.

CHAPTER 3 EXPERIMENTS:
Statistics for Describing, Exploring, and Comparing Data

3–1.   ***Comparing Heights of Men and Women*** In this experiment we use two small data sets as a quick introduction to using some of the basic Minitab features. (When beginning work with new software, it is wise to first work with small data sets so that they can be entered quickly if they are lost or damaged.) The data listed below are measured heights (in inches) of random samples of men and women (taken from Data Set 1 in Appendix B of the textbook).

Men	70.8	66.2	71.7	68.7	67.6	69.2
Women	64.3	66.4	62.3	62.3	59.6	63.6

a. Find the indicated characteristics of the heights of *men* and enter the results below.

Center: Mean: _____   Median: _____

Variation: St. Dev.: _____   Range: _____

5-Number Summary: Min.: _____ Q_1: _____ Q_2: _____ Q_3: _____ Max.: _____

Outliers: _____

b. Find the characteristics of the heights of *women* and enter the results below.

Center: Mean: _____   Median: _____

Variation: St. Dev.: _____   Range: _____

5-Number Summary: Min.: _____ Q_1: _____ Q_2: _____ Q_3: _____ Max.: _____

Outliers: _____

c. Compare the results from parts a and b.

_____

_____

_____

3–2.   ***Working with Larger Data Sets***   Repeat Experiment 3–1, but use the sample data for all 40 males and 40 females included in Data Set 1 that is found in Appendix B of the textbook. Instead of manually entering the 80 individual heights (which would be no fun at all), open the worksheets MHEALTH and FHEALTH that are found among the Minitab data sets on the CD–ROM that is included with the textbook.

a. Find the indicated characteristics of the heights of *men* and enter the results below.

Center: Mean: _____   Median: _____

Variation: St. Dev.:_____   Range: _____

5-Number Summary: Min.:_____ Q_1:_____ Q_2:_____ Q_3:_____   Max.:_____

Outliers: _____

b. Find the characteristics of the heights of *women* and enter the results below.

Center: Mean: _____   Median: _____

Variation: St. Dev.:_____   Range: _____

5-Number Summary: Min.:_____ Q_1:_____ Q_2:_____ Q_3:_____   Max.:_____

Outliers: _____

c. Compare the results from parts a and b.

_____

_____

_____

d. Are there are notable differences observed from the complete sets of sample data that could not be seen with the smaller samples listed in Experiment 3–1? If so, what are they?

_____

_____

_____

3–3.   ***Boxplots*** Use the same sets of data used in Experiment 3–2 and print boxplots for the heights of the 40 men and the heights of the 40 women. Include both boxplots in the same window so that they can be compared. Do the boxplots suggest any notable differences in the two sets of sample data?

_____

_____

Interpret the *asterisk* that appears in the Minitab display for the boxplot depicting the heights of the men.

_____

_____

3–4.   ***Effect of Outlier*** In this experiment we will study the effect of an *outlier*. Use the same heights of ten *men* used in Experiment 3-1, but change the first entry from 70.8 in. to 708 in. (This type of mistake often occurs when the key for the decimal point is not pressed with enough force.) The outlier of 708 in. is clearly a mistake, because a male with of height of 708 in. would be 59 feet tall, or about six stories tall. Although this outlier is a mistake, outliers are sometimes correct values that differ substantially from the other sample values.

Men: 708 66.2 71.7 68.7 67.6 69.2

Using this modified data set with the height of 70.8 in. changed to be the outlier of 708 in., find the following.

Center: Mean: _____   Median: _____

Variation: St. Dev.:_____   Range: _____

5-Number Summary: Min.:_____ Q_1:_____ Q_2:_____ Q_3:_____ Max.:_____

Outliers: _____

Based on a comparison of these results to those found in Experiment 3–1, how is the mean affected by the presence of an outlier?

_____

How is the median affected by the presence of an outlier?

_____

How is the standard deviation affected by the presence of an outlier?

_____

3–5. ***Sorting Data*** The ages of Oscar-winning actors are listed near the beginning of this chapter. *Sort* the data by arranging them in order from lowest to highest. List the first ten values in the sorted column.

_____

3–6. ***z Scores***   Retrieve the Minitab worksheet CANS.MTW, then use Minitab to find the *z* score corresponding to each of the 175 values for the cans that are 0.0109 in. thick. [Each value is an axial load, which is the weight (in pounds) that the can supports before being crushed.] Store the 175 *z* scores in column C3, then find the following.

Center: Mean: _____   Median: _____

Variation: St. Dev.:_____   Range: _____

5-Number Summary: Min.:_____ Q_1:_____ Q_2:_____ Q_3:_____ Max.:_____

Outliers: _____

What is notable about the above results? Specifically, what is notable about the value of the mean and standard deviation?

_____

_____

Will the same mean and standard deviation be obtained for *any* set of sample data? Explain how you arrived at your answer.

_____

_____

3–7. ***Combining Data*** Open the Minitab worksheet M&M, which consists of six different columns of data. Use **Data/Stack** (or **Manip/Stack** in earlier versions of Mintiab) to make a copy of the data all stacked together in column C7 of the current worksheet.

a. Find the following results for the combined data in column C7.

Center: Mean: _____   Median: _____

Variation: St. Dev.:_____   Range: _____

5-Number Summary: Min.:_____ Q_1:_____ Q_2:_____ Q_3:_____ Max.:_____

Outliers: _____

(*continued*)

b. Print a boxplot of the data set.

c. Describe the important characteristics of the data set. Be sure to address the nature of the distribution, measures of center, measures of variation, and any other important and notable features.

_____

_____

3–8. **_Statistics from a Dotplot_**   Shown below is a Minitab dotplot. Identify the values represented in this graph, enter them in column C1, then find the indicated results.

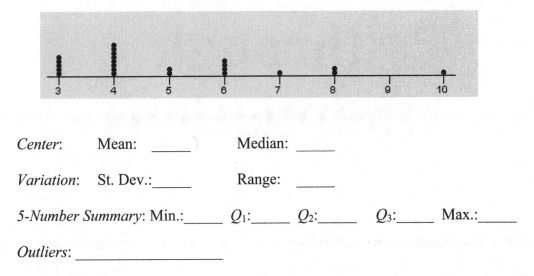

Center: Mean: _____   Median: _____

Variation: St. Dev.:_____   Range: _____

5-Number Summary: Min.:_____ Q_1:_____ Q_2:_____ Q_3:_____ Max.:_____

Outliers: _____

3–9. **_Comparing Data_** Open the Minitab worksheet COLA and use Minitab to compare the weights of regular Coke and the weights of diet Coke. Obtain printouts of relevant results. What do you conclude? Can you explain any substantial difference?

_____

_____

3–10. **_Working with Your Own Data_**   Through observation or experimentation, collect your own set of sample values. Obtain at least 40 values and try to select data from an interesting population. Use Minitab to explore the data. Obtain printouts of relevant results. Describe the nature of the data. That is, what do the values represent? Describe important characteristics of the data set, and include Minitab printouts to support your observations.

_____

_____

4

Probabilities through Simulations

4-1 Simulation Methods

The probability chapter in the textbook presents a variety of rules and methods for finding probabilities of different events. The textbook focuses on traditional approaches to computing probability values. This chapter in this manual/workbook focuses instead on an alternative approach based on *simulations*.

> A **simulation** of a procedure is a process that behaves the same way as the procedure, so that similar results are produced.

Mathematician Stanislaw Ulam once studied the problem of finding the probability of winning a game of solitaire, but the theoretical computations involved were too complicated. Instead, Ulam took the approach of programming a computer to simulate or "play" solitaire hundreds of times. The ratio of wins to total games played is the approximate probability he sought. This same type of reasoning was used to solve important problems that arose during World War II. There was a need to determine how far neutrons would penetrate different materials, and the method of solution required that the computer make various random selections in much the same way that it can randomly select the outcome of the rolling of a pair of dice. This neutron diffusion project was named the Monte Carlo Project and we now refer to general methods of simulating experiments as *Monte Carlo methods*. Such methods are the focus of this chapter. The concept of simulation is quite easy to understand with simple examples.

- We could simulate the rolling of a die by using Minitab to randomly generate whole numbers between 1 and 6 inclusive, provided that the computer selects from the numbers 1, 2, 3, 4, 5, and 6 in such a way that those outcomes are equally likely.
- We could simulate births by flipping a coin, where "heads" represents a baby girl and "tails" represents a baby boy. We could also simulate births by using Minitab to randomly generate 1s (for baby girls) and 0s (for baby boys).

It is extremely important to construct a simulation so that it behaves just like the real procedure. The following example illustrates a right way and a wrong way.

EXAMPLE Describe a procedure for simulating the rolling of a pair of dice.

SOLUTION In the procedure of rolling a pair of dice, each of the two dice yields a number between 1 and 6 (inclusive) and those two numbers are then added. Any simulation should do exactly the same thing.

> *Right way to simulate rolling two dice:* Randomly generate one number between 1 and 6, then randomly generate another number between 1 and 6, then add the two results.
>
> *Wrong way to simulate rolling two dice:* Randomly generate numbers between 2 and 12. This procedure is similar to rolling dice in the sense that the results are always between 2 and 12, but these outcomes between 2 and 12 are equally likely. With real dice, the values between 2 and 12 are *not* equally likely. This simulation would yield terrible results.

4-2   Minitab Simulation Tools

Minitab includes several different tools that can be used for simulations.

1.   Click on the main menu item of **Calc**.

2.   Click on the subdirectory item of **Random Data**.

You will get a long list of different options, but we will use four of them (integer, uniform, Bernouli, normal) that are particularly relevant to the simulation techniques of this chapter.

- **Integer:** Generates a sample of randomly selected integers with a specified minimum and maximum. This Minitab tool is particularly good for a variety of different simulations. Shown below is the dialog box that is provided when you select **Calc/Random Data/ Integer**. The entries in this box tell Minitab to randomly generate 500 values between 1 and 6 inclusive, and put them in column C1. This simulates rolling a single die 500 times.

- **Uniform:** Select **Calc**, then **Random Data**, then **Uniform.** The generated values are "uniform" in the sense that all possible values have the same chance of being selected. The Minitab dialog box shown below has entries indicating that 500 values should be randomly selected between 0.0 and 9.0. If you want whole numbers, use the Integer distribution instead of the Uniform distribution.

- **Bernouli:** The dialog box below has entries that result in a column of 0s (failure) and 1s (success), where the probability of success is 0.5. This simulates flipping a coin 500 times. The probability can be changed for different circumstances. The number of values and the probability can be changed as desired.

- **Normal:** Select **Calc**, then **Random Data**, then **Normal** to generate a sample of data randomly selected from a population having a normal distribution. (Normal distributions are discussed in Chapter 6. For now, think of a normal distribution as one that is bell–shaped.) You must enter the population mean and standard deviation. The entries in the dialog box below result in a simulated sample of 500 IQ scores taken from a population with a bell-shaped distribution that has a mean of 100 and a standard deviation of 15. Shown below the dialog box is a Minitab-generated histogram displaying the bell-shaped nature of the generated data.

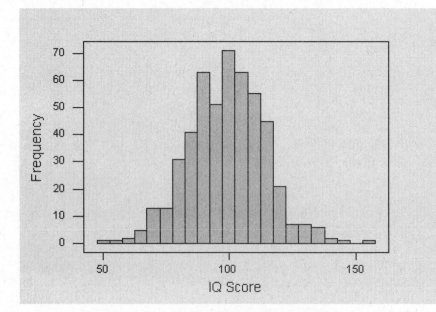

4-3 Simulation Examples

We will now illustrate the preceding Minitab features by describing specific simulations.

Simulation 1: Generating 50 births (boys/girls)

To simulate 50 births with the assumption that boys and girls are equally likely, use either of the following:

- Use Minitab's **Integer** distribution (see Section 4-2 of this manual/workbook) to generate 50 integers between 0 and 1. If you arrange the results in order (using the **Sort** feature as described in Section 3-4), it is very easy to count the number of 0s (or boys) and the number of 1s (or girls).

- Use Minitab's **Bernouli** distribution. Enter 50 for the sample size and enter 0.5 for the probability. Again, it is very easy to count the number of 0s (or boys) and the number of 1s (or girls) if the data are sorted, as described in Section 3-4 of this manual/workbook.

Simulation 2: Rolling a *single* die 60 times

To simulate 60 rolls of a single die, use Minitab's **Integer** distribution to generate 60 integers between 1 and 6. Again, arranging them in order makes it easy to count the number of 1s, 2s, and so on.

Simulation 3: Rolling a *pair* of dice 750 times

To simulate the rolling of a *pair* of dice 750 times, use column C1 for one die, use column C2 for the second die, then let column C3 be the *sum* of columns C1 and C2. That is, create column C1 by using the method described in Simulation 2 above. Then create column C2 by again using the method described in Simulation 2 above. Then use **Calc/Calculator** to enter the expression C1 + C2 with the result stored in column C3.

Simulation 4: Generating 25 birth dates

Instead of generating 25 results such as "January 1," or "November 27," randomly generate 25 integers between 1 and 365 inclusive. (We are ignoring leap years). Use Minitab's **Integer** distribution and enter 25 for the sample size. Also enter 1 for the minimum and enter 365 for the maximum. If you sort the simulated birth dates by using the procedure in Section 3-4, it becomes easy to scan the sorted list and determine whether there are two birth dates that are the same. If there are two birth dates that are the same, they will show up as *consecutive* equal values in the sorted list.

CHAPTER 4 EXPERIMENTS: Probabilities Through Simulations

4—1.   ***Birth Simulation***   Use Minitab to simulate 500 births, where each birth results in a boy or girl. Sort the results, count the number of girls, and enter that value here:_____

Based on that result, estimate the probability of getting a girl when a baby is born. Enter the estimated probability here: _____

The preceding estimated probability is likely to be different from 0.5. Does this suggest that the computer's random number generator is defective? Why or why not?

_____

_____

4—2.   ***Dice Simulation***   Use Minitab to simulate 1000 rolls of a pair of dice. Sort the results, then find the number of times that the total was exactly 7. Enter that value here:_____

Based on that result, estimate the probability of getting a 7 when two dice are rolled. Enter the estimated probability here:_____

How does this estimated probability compare to the computed (theoretical) probability of 0.167? _____

_____

4—3.   ***Probability of Exactly 11 Girls***
   a.   Use Minitab to simulate 20 births. Does the result consist of exactly 11 girls? _____
   b.   Repeat part (a) nine more times and record the result from part (a) along with the other nine results here: ___ ___ ___ ___ ___ ___ ___ ___ ___ ___
   c.   Based on the results from part (b), what is the estimated probability of getting exactly 11 girls in 20 births? _____

4—4.   ***Probability of at Least 11 Girls***
   a.   Use Minitab to simulate 20 births. Does the result consist of at least 11 girls? _____
   b.   Repeat part (a) nine more times and record the result from part (a) along with the other nine results here: ___ ___ ___ ___ ___ ___ ___ ___ ___ ___
   c.   Based on the results from part (b), what is the estimated probability of getting at least 11 girls in 20 births? _____

4—5.   ***Simulating Motorcycle Drivers***   In a study of fatalities caused by motorcycle crashes, it was found that 95% of motorcycle drivers are men (based on data from "Motorcycle Rider Conspicuity and Crash Related Injury," by Wells et al, *BJM USA*). Simulate the random selection of 20 motorcycle drivers. Each individual outcome should consist of an indication of whether the motorcycle driver is a man or woman. Enter the genders here:

_____

4—6.   *Guessing Simulation* Use Minitab to conduct a simulation that can be used to estimate the probability of getting at least six correct responses when random guesses are made for all 10 multiple choice questions in a quiz. Each question has five possible answers (a, b, c, d, e) and only one of them is correct. Enter the estimated probability here: _____ Describe the simulation process that was used.

_____

4—7.   *Probability of at Least 55 Girls* Use Minitab to conduct a simulation for estimating the probability of getting at least 55 girls in 100 births. Enter the estimated probability here:_____   Describe the procedure used to obtain the estimated probability.

_____

In testing a gender-selection method, assume that the Biogene Technology Corporation conducted an experiment with 100 couples who were treated, and that the 100 births included at least 55 girls. What should you conclude about the effectiveness of the treatment?

_____

4—8.   *Probability of at Least 65 Girls* Use Minitab to conduct a simulation for estimating the probability of getting at least 65 girls in 100 births. Enter the estimated probability here:_____
Describe the procedure used to obtain the estimated probability.

_____

In testing a gender-selection method, if the Biogene Technology Corporation conducted an experiment with 100 couples who were treated, and the 100 births included at least 65 girls, what should you conclude about the effectiveness of the treatment?

_____

4—9.   *Simulating Families of Five Children* Develop a simulation for finding the probability of getting at least two girls in a family of five children. Simulate 100 families. Describe the simulation, then estimate the probability based on its results.

_____

4—10. *Simulating Three Dice* Develop a simulation for rolling three dice. Simulate the rolling of the three dice 100 times. Describe the simulation, then use it to estimate the probability of getting a total of 10 when three dice are rolled.

_____

_____

4–11. *Simulating Left–Handedness* Ten percent of us are left-handed. In a study of dexterity, people are randomly selected in groups of five. Develop a simulation for finding the probability of getting at least one left-handed person in a group of five. Simulate 100 groups of five. How does the probability compare to the correct result of 0.410, which can be found by using the probability rules in the textbook?

_____

_____

4–12. *Simulating Hybridization* When Mendel conducted his famous hybridization experiments, he used peas with green pods and yellow pods. One experiment involved crossing peas in such a way that 25% of the offspring peas were expected to have yellow pods. Develop a simulation for finding the probability that when two offspring peas are produced, at least one of them has yellow pods. Generate 100 pairs of offspring. How does the result compare to the correct probability of 7/16, which can be found by using the probability rules in the textbook?

_____

_____

4–13. *Effectiveness of Drug* It has been found that when someone tries to stop smoking under certain circumstances, the success rate is 20%. A new nicotine substitute drug has been designed to help those who wish to stop smoking. In a trial of 50 smokers who use the drug while trying to stop, it was found that 12 successfully stopped. The drug manufacturer argues that the 12 successes are better than the 10 that would be expected without the drug, so the drug is effective. Conduct a simulation of 50 smokers trying to stop, and assume that the drug has no effect, so the success rate continues to be 20%. Repeat the simulation several times and determine whether 12 successes could easily occur with an ineffective drug. What do you conclude about the effectiveness of the drug?

_____

_____

4–14. *Birthdays* Simulate a class of 25 birth dates by randomly generating 25 integers between 1 and 365. (We will ignore leap years.) Arrange the birth dates in ascending order, then examine the list to determine whether at least two birth dates are the same. (This is easy to do, because any two equal integers must be next to each other.)

Generated "birth dates:"   ___ ___ ___ ___ ___ ___ ___ ___ ___ ___ ___ ___ ___
                             ___ ___ ___ ___ ___ ___ ___ ___ ___ ___ ___ ___

Are at least two of the "birth dates" the same? _____

4—15. *Birthdays* Repeat the preceding experiment nine additional times and record all ten of the yes/no responses here:

____ ____ ____ ____ ____ ____ ____ ____ ____ ____

Based on these results, what is the probability of getting at least two birth dates that are the same (when a class of 25 students is randomly selected)? _____

4—16. *Birthdays* Repeat Experiments 4-14 and 4-15 for 50 people instead of 25. Based on the results, what is the estimated probability of getting at least two birth dates that are the same (when a class of 50 students is randomly selected)? _____

4—17. *Birthdays* Repeat Experiments 4-14 and 4-15 for 100 people instead of 25. Based on the results, what is the estimated probability of getting at least two birth dates that are the same (when a class of 100 students is randomly selected)? _____

4—18. *Normally Distributed Heights* Simulate 1000 heights of adult women. (Adult women have normally distributed heights with a mean of 63.6 in. and a standard deviation of 2.5 in.) Arrange the data in ascending order, then examine the results and estimate the probability of a randomly selected woman having a height between 64.5 in. and 72 in. (Those were the height restrictions for women to fit into Russian Soyuz spacecraft when NASA and Russia ran joint missions.) Enter the estimated probability here: _____

4—19. *Normally Distributed Bulb Lives* The lifetimes of 75-watt light bulbs manufactured by the Lectrolyte Company have a mean of 1000 hours and a standard deviation of 100 hours. Generate a normally distributed sample of 500 bulb lifetimes by using the given mean and standard deviation. Examine the sorted results to estimate the probability of randomly selecting a light bulb that lasts between 850 hours and 1150 hours. Enter the result here. _____

4—20. *Normally Distributed IQ Scores* IQ scores are normally distributed with a mean of 100 and a standard deviation of 15. Generate a normally distributed sample of 800 IQ scores by using the given mean and standard deviation. Sort the results (arrange them in ascending order).

a. Examine the sorted results to estimate the probability of randomly selecting someone with an IQ score between 90 and 110 inclusive. Enter the result here. _____

b. Examine the sorted results to estimate the probability of randomly selecting someone with an IQ score greater than 115. _____

c. Examine the sorted results to estimate the probability of randomly selecting someone with an IQ score less than 120. _____

d. Repeat part a of this experiment nine more times and list all ten probabilities here.

____ ____ ____ ____ ____ ____ ____ ____ ____ ____

(continued)

e. Examine the ten probabilities obtained above and comment on the *consistency* of the results.

_____

f. How might we modify this experiment so that the results can become more consistent?

_____

g. If the results appear to be very consistent, what does that imply about any individual sample result?

_____

4—21. *Law of Large Numbers*   In this experiment we test the Law of Large Numbers, which states that "as an experiment is repeated again and again, the empirical probability of success tends to approach the actual probability." We will use a simulation of a single die, and we will consider a success to be the outcome of 1. (Based on the classical definition of probability, we know that $P(1) = 1/6 = 0.167$.)

a. Simulate 5 trials by generating 5 integers between 1 and 6. Count the number of 6s that occurred and divide that number by 5 to get the empirical probability.
Based on 5 trials, $P(1) =$ _____.

b. Repeat part (a) for 25 trials. Based on 25 trials, $P(1)=$ _____.

c. Repeat part (a) for 50 trials. Based on 50 trials, $P(1)=$ _____.

d. Repeat part (a) for 500 trials. Based on 500 trials, $P(1)=$ _____.

e. Repeat part (a) for 1000 trials. Based on 1000 trials, $P(1) =$ _____.

f. In your own words, generalize these results in a restatement of the Law of Large Numbers.

_____

_____

_____

_____

4—22. ***Sticky Probability Problem***   Consider the following exercise, which is extremely difficult to solve with the formal rules of probability.

> *Two points along a straight stick are randomly selected. The stick is then broken at these two points. Find the probability that the three pieces can be arranged to form a triangle.*

Instead of attempting a solution using formal rules of probability, we will use a simulation. The length of the stick is irrelevant, so assume it's one unit long and its length is measured from 0 at one end to 1 at the other end. Use Minitab to randomly select the two break points with the random generation of two numbers from a uniform distribution with a minimum of 0, a maximum of 1, and 4 decimal places. Plot the break points on the "stick" below.

0 _____ 1

A triangle can be formed if the longest segment is less than 0.5, so enter the lengths of the three pieces here: _____ _____ _____

Can a triangle be formed?

Now repeat this process nine more times and summarize all of the results below.

Trial	Break Points	Triangle formed?
1		
2		
3		
4		
5		
6		
7		
8		
9		
10		

Based on the ten trials, what is the estimated probability that a triangle can be formed? _____ This estimate gets better with more trials.

5

Probability Distributions

5-1 Exploring Probability Distributions

The textbook includes a chapter on Probability Distributions, and its focus is *discrete* probability distributions only. These important definitions are introduced:

Definitions

A **random variable** is a variable (typically represented by x) that has a single numerical value, determined by chance, for each outcome of a procedure.

A **probability distribution** is a graph, table, or formula that gives the probability for each value of the random variable.

When working with a probability distribution, we should consider the same important characteristics introduced in Chapter 2:

1. **Center:** Measure of center, which is a representative or average value that gives us an indication of where the middle of the data set is located

2. **Variation:** A measure of the amount that the values vary among themselves

3. **Distribution:** The nature or shape of the distribution of the data, such as bell-shaped, uniform, or skewed

4. **Outliers:** Sample values that are very far away from the vast majority of the other sample values

5. **Time:** Changing characteristics of the data over time

Section 5-1 of this manual/workbook addresses these important characteristics for probability distributions. The characteristics of center and variation are addressed with formulas for finding the mean, standard deviation, and variance of a probability distribution. The characteristic of distribution is addressed through the graph of a probability histogram.

Although Minitab is not designed to deal directly with a probability distribution, it can often be used. Let's consider past results from baseball's World Series found in the *Information Please Almanac*. There is a 0.1818 probability that a baseball World Series contest will last four games, a 0.2121 probability that it will last five games, a 0.2323 probability that it will last six games, and a 0.3737 probability that it will last seven games. The probability distribution is summarized in the table at the right.

x	$P(x)$
4	0.1818
5	0.2121
6	0.2323
7	0.3737

If you examine the data in the table, you can verify that a probability distribution is defined because the two key requirements are satisfied: (1) The sum of the probabilities is equal to 1; (2) each individual probability is between 0 and 1 inclusive. (The sum of the probabilities is actually 0.9999, but it can be considered to be 1 when rounding errors are taken into account.)

Having determined that the above table does define a probability distribution, let's now see how we can use Minitab to find the mean μ and standard deviation σ. Although Minitab is not designed to directly calculate the value of the mean, there is a way to calculate exact values, and there is also a way to get very good estimated values.

Finding Exact Values of μ and σ for a Probability Distribution

1. Enter the values of the random variable x in column C1.

2. Enter the corresponding probabilities in column C2.

3. Calculate the mean μ by using this formula in the textbook for the mean of a probability distribution: $\mu = \sum[x \cdot P(x)]$. This is accomplished by clicking on **Calc**, selecting **Calculator**, and entering the expression sum(C1*C2) with the result stored in column C3. Column C3 will store the value of the mean. See the Minitab screen display below.

4. Now proceed to calculate the value of the standard deviation σ by using this formula from the textbook: $\sigma = \sqrt{\left[\Sigma x^2 \cdot P(x)\right] - \mu^2}$. This is accomplished by selecting **Calc**, then **Calculator**, and entering the expression

$$\text{sqrt}(\text{sum}(C1**2*C2) - C3**2)$$

with the result stored in column C4.

As an example, if you use the above procedure with the probability distribution defined by the table on page 53, the result will be as shown below. We can see that $\mu = 5.7974$ (or 5.8 after rounding) and the standard deviation is $\sigma = 1.12940$ (or 1.1 after rounding). (The names of the columns in the display below were manually entered.)

↓	C1	C2	C3	C4
	x	P(x)	Mean	St.Dev.
1	4	0.1818	5.7974	1.12940
2	5	0.2121		
3	6	0.2323		
4	7	0.3737		

Estimating Values of μ and σ for a Probability Distribution

The above procedure for exact results is messy and somewhat difficult to remember. An easier approach for finding the mean μ and standard deviation σ is to collect a large random sample from the probability distribution, then calculate the mean and standard deviation by using **Stat**, then **Basis Statistics**, then **Display Descriptive Statistics**. Here is the Minitab procedure.

1. Enter the values of the random variable x in column C1.

2. Enter the corresponding probabilities in column C2.

3. Select **Calc**, then **Random Data**, then **Discrete**.

4. You should now get a dialog box like the one shown on the next page. The entries in this dialog box tell Minitab to generate 1,000,000 sample values randomly selected from the probability distribution defined by the values of x and $P(x)$ in columns C1 and C2.

5. Using the sample values in column C3, find the mean and standard deviation by using **Stat/Basic Statistics/Display Descriptive Statistics**.

As an example, enter the values of x and $P(x)$ listed in the table on page 53. Using the above procedure, the author found a mean of 5.8006 (while the exact value is 5.7974) and a standard deviation of 1.1275 (while the exact value is 1.12940). The simulated sample provided results that are very close to the exact values.

Discrete Distribution ☒

Generate `1000000`   rows of data

Store in column(s):

C3

Values in: C1

Probabilities in: C2

Select

Help   OK   Cancel

5-2 Binomial Distributions

A **binomial distribution** is defined in the textbook to be to be a probability distribution that meets all of the following requirements:

1. The experiment must have a fixed number of trials.
2. The trials must be independent. (The outcome of any individual trial doesn't affect the probabilities in the other trials.)
3. Each trial must have all outcomes classified into two categories.
4. The probabilities must remain constant for each trial.

We also introduced notation with S and F denoting success and failure for the two possible categories of all outcomes. Also, p and q denote the probabilities of S and F, respectively, so that $P(S) = p$ and $P(F) = q$. We also use the following symbols.

n denotes the fixed number of trials

x denotes a specific number of successes in n trials so that x can be any whole number between 0 and n, inclusive.

p denotes the probability of success in *one* of the n trials.

q denotes the probability of failure in *one* of the n trials.

$P(x)$ denotes the probability of getting exactly x successes among the n trials.

The textbook describes three methods for determining probabilities in binomial experiments. Method 1 uses the binomial probability formula:

$$P(x) = \frac{n!}{(n - x)!x!} p^x q^{n-x}$$

Method 2 requires use of Table A-1, the table of binomial probabilities. Method 3 requires computer usage. We noted in the textbook that if a computer and software are available, this third method of finding binomial probabilities is fast and easy, as shown in the following Minitab procedure.

Minitab Procedure for Finding Probabilities with a Binomial Distribution

1. In column C1, enter the values of the random variable x for which you want probabilities. (You can enter all values of x as 0, 1, 2, . . .)

2. Click on **Calc** from the main menu.

3. Select **Probability Distributions**.

4. Select **Binomial**.

5. The entries in this dialog box shown below correspond to a binomial distribution with 4 trials, and a probability of success equal to 0.2. The probability values are to be computed for the values of x that have been entered in column C1, and the results will be stored in column C2.

Consider the following example.

> ***Analysis of Multiple Choice Answers***   Use the binomial probability formula to find the probability of getting exactly 3 correct answers when random guesses are made for 4 multiple choice questions. That is, find the value $P(3)$ given that $n = 4$, $x = 3$, $p = 0.2$, and $q = 0.8$.

Begin by entering the values of 0, 1, 2, 3, and 4 in column C1, and continue with the above procedure. Columns C1 and C2 will display the values shown below. We can easily see from this display that $P(3) = 0.0256$, which is the same value obtained by using the binomial probability formula. (*Caution*: Don't use the numbers in the column at the extreme left for the values of x; be sure to use the values in column C1 for the values of x.) The table of binomial probabilities (Table A-1) in Appendix B of the textbook includes limited values of n and p, but Minitab is much more flexible in the values of n and p that can be used.

↓	C1	C2
1	0	0.4096
2	1	0.4096
3	2	0.1536
4	3	0.0256
5	4	0.0016

5-3 Poisson Distributions

See textbooks in the Triola Statistics Series for a discussion of the Poisson distribution. (This discussion is not included in *Essentials of Statistics*.) A Poisson distribution is a discrete probability distribution that applies to occurrences of some event *over a specified interval*. The random variable x is the number of occurrences of the event in an interval, such as time, distance, area, volume, or some similar unit. The probability of the event occurring x times over an interval is given by this formula:

$$P(x) = \frac{\mu^x \cdot e^{-\mu}}{x!} \qquad \text{where } e \approx 2.71828$$

We also noted that the Poisson distribution is sometimes used to approximate the binomial distribution when $n \geq 100$ and $np \leq 10$; in such cases, we use $\mu = np$. If using Minitab, the Poisson approximation to the binomial distribution isn't used as often, because we can easily find binomial probabilities for a wide range of values for n and p.

Minitab Procedure for Finding Probabilities for a Poisson Distribution

1.   Determine the value of the mean μ.

2.   In column C1, enter the values of the random variable x for which you want probabilities. (You can enter values of x as 0, 1, 2, ...)

3.   Click on **Calc** from the main menu.

4. Select **Probability Distributions**.

5. Select **Poisson**.

6. You will now see a dialog box like the one shown below. The probability values are to be computed for the values of x that have been entered in column C1, and the results will be stored in column C2.

Consider this example:

> ***World War II Bombs*** In analyzing hits by V-1 buzz bombs in World War II, South London was subdivided into 576 regions, each with an area of 0.25 km^2. A total of 535 bombs hit the combined area of 576 regions. If a region is randomly selected, find the probability that it was hit exactly twice.

Because a total of 535 bombs hit 576 regions, the mean number of hits is 535/576 = 0.929. Having found the required mean μ, we can now proceed to use Minitab as described above. If you enter the x values of 0, 1, 2, 3, 4, and 5 in column C1, you will results in columns C1 and C2 showing that $P(2) = 0.170428$, which is the probability that a region would be hit exactly two times. (Again, be careful to refer to column C1 for the values of x.)

5-4 Cumulative Probabilities

The main objective of this section is to reinforce the point that cumulative probabilities are often critically important. By *cumulative* probability, we mean the probability that the random variable *x* has a range of values instead of a single value. Here are typical examples:

- Find the probability of getting *at least* 13 girls in 14 births.
- Find the probability of *more than* 5 wins when roulette is played 200 times.

The textbook stresses that in many cases, a *cumulative* probability is much more important than the probability of any individual event. The textbook includes criteria for determining when results are unusual, and note that the following criteria involve cumulative probabilities.

Using Probabilities to Determine When Results Are Unusual

- **Unusually high number of successes:** *x* successes among *n* trials is an *unusually high* number of successes if $P(x \text{ or more}) \leq 0.05^*$.

- **Unusually low number of successes:** *x* successes among *n* trials is an *unusually low* number of successes if $P(x \text{ or fewer}) \leq 0.05^*$.

  *The value of 0.05 is commonly used, but is not absolutely rigid. Other values, such as 0.01, could be used to distinguish between events that can easily occur by chance and events that are very unlikely to occur by chance.

Consider a test of the MicroSort gender selection technique. In one early trial, there were 13 girls among 14 babies. Is this result unusual? Does this result really suggest that the technique is effective, or could it be that there were 13 girls among 14 babies just by chance? It was noted that in answering this key question, the relevant probability is the *cumulative* probability of getting 13 or more girls, not the probability of getting exactly 13 girls. The textbook supported that point with an example, reproduced here because it is so important:

> *Suppose you were flipping a coin to determine whether it favors heads, and suppose 1000 tosses resulted in 501 heads. This is not evidence that the coin favors heads, because it is very easy to get a result like 501 heads in 1000 tosses just by chance. Yet, the probability of getting exactly 501 heads in 1000 tosses is actually quite small: 0.0252. This low probability reflects the fact that with 1000 tosses, any specific number of heads will have a very low probability. However, we do not consider 501 heads among 1000 tosses to be unusual, because the probability of getting at least 501 heads is high: 0.487.*

Cumulative probabilities therefore play a critical role in identifying results that are considered to be *unusual*. Later chapters focus on this important concept. If you examine the dialog boxes used for finding binomial probabilities and Poisson probabilities, you can see that in addition to obtaining probabilities for specific numbers of successes, you can also obtain *cumulative* probabilities.

CHAPTER 5 EXPERIMENTS: Probability Distributions

5-1.   A probability distribution is described below. Use Minitab with the procedure described in Section 5-1 of this manual/workbook for the following.

   a.   Using the *exact* procedure, find the mean and standard deviation.
         Mean:_____   St. Dev.:_____
   b.   Using the *approximate* procedure, estimate the mean and standard deviation.
         Mean:_____   St. Dev.:_____
   c.   Compare the exact values to the estimated values.

_____

Genetic Disorder *Three males with an X–linked genetic disorder have one child each. The random variable x is the number of children among the three who inherit the X–linked genetic disorder, and the probability distribution for that number of children is given in the table below.*

x	$P(x)$
0	0.4219
1	0.4219
2	0.1406
3	0.0156

5-2.   A probability distribution is described below. Use Minitab with the procedure described in Section 5-1 of this manual/workbook for the following.

   a.   Using the *exact* procedure, find the mean and standard deviation.
         Mean:_____   St. Dev.:_____
   b.   Using the *approximate* procedure, estimate the mean and standard deviation.
         Mean:_____   St. Dev.:_____
   c.   Compare the exact values to the estimated values.

_____

Life Insurance *The Telektronic Company provides life insurance policies for its top four executives, and the random variable x is the number of those employees who live through the next year.*

x	$P(x)$
0	0.0000
1	0.0001
2	0.0006
3	0.0387
4	0.9606

5-3.   A probability distribution is described below. Use Minitab with the procedure described
        in Section 5-1 of this manual/workbook for the following.
        a.   Using the *exact* procedure, find the mean and standard deviation.
                Mean:_____ St. Dev.:_____
        b.   Using the *approximate* procedure, estimate the mean and standard deviation.
                Mean:_____ St. Dev.:_____
        c.   Compare the exact values to the estimated values.

Genetics Experiment *A genetics experiment involves offspring peas in groups of four.
A researcher reports that for one group, the number of peas with white flowers has a
probability distribution as given in the accompanying table.*

x	$P(x)$
0	0.04
1	0.16
2	0.80
3	0.16
4	0.04

5-4.   *Binomial Probabilities*   Use Minitab to find the binomial probabilities corresponding to
        $n = 4$ and $p = 0.05$. Enter the results below, along with the corresponding results found in
        Table A-1 of the textbook.

x	$P(x)$ from Minitab	$P(x)$ from Table A-1
0		
1		
2		
3		
4		

        By comparing the above results, what advantage does Minitab have over Table A-1?

5-5.   *Binomial Probabilities*   Consider the binomial probabilities corresponding to $n = 4$ and
        $p = 1/4$ (or 0.25). If you attempt to use Table A-1 for finding the probabilities, you will
        find that the table does not apply to a probability of $p = 1/4$. Use Minitab to find the
        probabilities and enter the results below.

        $P(0) =$ _____ $P(1) =$ _____ $P(2) =$ _____ $P(3) =$ _____ $P(4) =$ _____

5-6. ***Binomial Probabilities*** Assume that boys and girls are equally likely and 100 births are randomly selected. Use Minitab with $n = 100$ and $p = 0.5$ to find $P(x)$, where x represents the number of girls among the 100 babies.

   a. $P(35) =$ _____

   b. $P(45) =$ _____

   c. $P(50) =$ _____

5-7. ***Binomial Probabilities*** We often assume that boys and girls are equally likely, but the actual values are $P(\text{boy}) = 0.5121$ and $P(\text{girl}) = 0.4879$. Repeat Experiment 5-6 using these values, then compare these results to those obtained in Experiment 5-6.

   a. $P(35) =$ _____ _____

   b. $P(45) =$ _____ _____

   c. $P(50) =$ _____ _____

5-8. ***Cumulative Probabilities*** Assume that $P(\text{boy}) = 0.5121$, $P(\text{girl}) = 0.4879$, and that 100 births are randomly selected. Use Minitab to find the probability that the number of girls among 100 babies is . . .

   a. Fewer than 60   _____

   b. Fewer than 48   _____

   c. At most 30   _____

   d. At least 55   _____

   e. More than 40   _____

5-9. ***Identifying 0+*** In Table A-1 from the textbook, the probability corresponding to $n = 12$, $p = 0.10$, and $x = 6$ is shown as 0+. Use Minitab to find the corresponding probability and enter the result here. _____

5-10. ***Identifying 0+*** In Table A-1 from the textbook, the probability corresponding to $n = 15$, $p = 0.80$, and $x = 5$ is shown as 0+. Use Minitab to find the corresponding probability and enter the result here. _____

5-11. ***Identifying a Probability Distribution*** Use Minitab to construct a table of x and $P(x)$ values corresponding to a binomial distribution in which $n = 26$ and $p = 0.3$. Enter the table in the margin.

Exercises 5–12 through 5-16 involve binomial distributions. Use Minitab for those exercises.

5–12. *Color Blindness* Nine percent of men and 0.25% of women cannot distinguish between the colors red and green. This is the type of color blindness that causes problems with traffic signals. If six men are randomly selected for a study of traffic signal perceptions, find the probability that exactly two of them cannot distinguish between red and green. Enter the probability here: _____

5–13. *Acceptance Sampling*   The Telektronic Company purchases large shipments of fluorescent bulbs and uses this acceptance sampling plan: Randomly select and test 24 bulbs, then accept the whole batch if there is only one or none that doesn't work. If a particular shipment of thousands of bulbs actually has a 4% rate of defects, what is the probability that this whole shipment will be accepted? _____

5–14. *IRS Audits* The Hemingway Financial Company prepares tax returns for individuals. (Motto: "We also write great fiction.") According to the Internal Revenue Service, individuals making $25,000–$50,000 are audited at a rate of 1%. The Hemingway Company prepares five tax returns for individuals in that tax bracket, and three of them are audited.
   a.   Find the probability that when 5 people making $25,000–$50,000 are randomly selected, exactly 3 of them are audited. _____
   b.   Find the probability that at least three are audited. _____
   c.   Based on the preceding results, what can you conclude about the Hemingway customers? Are they just unlucky, or are they being targeted for audits?

   _____

5–15. *Overbooking Flights* Air America has a policy of booking as many as 15 persons on an airplane that can seat only 14. (Past studies have revealed that only 85% of the booked passengers actually arrive for the flight.) Find the probability that if Air America books 15 persons, not enough seats will be available. Is this probability low enough so that overbooking is not a real concern for passengers? _____

5–16. *Drug Reaction* In a clinical test of the drug Viagra, it was found that 4% of those in a placebo group experienced headaches.
   a.   Assuming that the same 4% rate applies to those taking Viagra, find the probability that among 8 Viagra users, 3 experience headaches. _____
   b.   Assuming that the same 4% rate applies to those taking Viagra, find the probability that among 8 randomly selected users of Viagra, all 8 experienced a headache. _____
   c.   If all 8 Viagra users were to experience a headache, would it appear that the headache rate for Viagra users is different than the 4% rate for those in the placebo group? Explain. _____

Exercises 5-17 through 5-20 involve Poisson distributions. Use Minitab for those exercises.

5–17.   ***Radioactive Decay*** Radioactive atoms are unstable because they have too much energy. When they release their extra energy, they are said to decay. When studying Cesium 137, it is found that during the course of decay over 365 days, 1,000,000 radioactive atoms are reduced to 977,287 radioactive atoms.
   a.   Find the mean number of radioactive atoms lost through decay in a day. _____
   b.   Find the probability that on a given day, 50 radioactive atoms decayed.   _____

5-18.   ***Aircraft Hijackings*** For the past few years, there has been a yearly average of 29 aircraft hijackings worldwide (based on data from the FAA). The mean number of hijackings per day is estimated as $\mu = 29/365$. If the United Nations is organizing a single international hijacking response team, there is a need to know about the chances of multiple hijackings in one day. Find the probability that the number of hijackings (x) in one day is 0 or 1.
_____
What do you conclude about the United Nation's organizing of a single response team?

_____

5-19.   ***Deaths From Horse Kicks*** A classic example of the Poisson distribution involves the number of deaths caused by horse kicks of men in the Prussian Army between 1875 and 1894. Data for 14 corps were combined for the 20-year period, and the 280 corps-years included a total of 196 deaths. After finding the mean number of deaths per corps-year, find the probability that a randomly selected corps-year has the following numbers of deaths.
   a. 0 _____   b. 1 _____   c. 2 _____   d. 3 _____   e. 4 _____
   The actual results consisted of these frequencies: 0 deaths (in 144 corps-years); 1 death (in 91 corps-years); 2 deaths (in 32 corps-years); 3 deaths (in 11 corps-years); 4 deaths (in 2 corps-years). Compare the actual results to those expected from the Poisson probabilities. Does the Poisson distribution serve as a good device for predicting the actual results?

_____

5-20.   ***Homicide Deaths*** In one year, there were 116 homicide deaths in Richmond, Virginia (based on "A Classroom Note On the Poisson Distribution: A Model for Homicidal Deaths In Richmond, VA for 1991," *Mathematics and Computer Education,* by Winston A. Richards). For a randomly selected day, find the probability that the number of homicide deaths is
   a. 0 _____   b. 1 _____   c. 2 _____   d. 3 _____   e. 4 _____

Compare the calculated probabilities to these actual results: 268 days (no homicides); 79 days (1 homicide); 17 days (2 homicides); 1 day (3 homicides); there were no days with more than 3 homicides._____

_____

6

Normal Distributions

6-1 Simulating and Generating Normal Data

We can learn much about the behavior of normal distributions by analyzing samples obtained from them. Sampling from real populations is often time consuming and expensive, but we can use the wonderful power of computers to obtain samples from theoretical normal distributions, and Minitab has such a capability, as described below. Let's consider IQ scores. IQ tests are designed to produce a mean of 100 and a standard deviation of 15, and we expect that such scores are normally distributed. Suppose we want to learn about the variation of sample means for samples of IQ scores. Instead of going out into the world and randomly selecting groups of people, we can sample from theoretical populations. We can then learn much about the distribution of sample means. The following procedure allows you to obtain a random sample from a normally distributed population with a given mean and standard deviation.

Generating a Random Sample from a Normally Distributed Population

1. Click on the main menu item of **Calc**.
2. Click on **Random Data**.
3. Click on **Normal**.
4. You will now get a dialog box such as the one shown below. You must enter the population mean and standard deviation. The entries in the dialog box below result in simulated sample of 500 IQ scores taken from a population with a normal distribution that has a mean of 100 and a standard deviation of 15. The sample values are stored in column C1.

Important: *When using the above procedure, the generated sample will not have a mean equal to the specified population mean, and the standard deviation of the sample will not have a value equal to the specified standard deviation.* When the author generated 500 sample values with the above dialog box, he obtained sample values with a mean of 100.28 and a standard deviation of 13.98. Remember, you are sampling from a population with the specified parameters; you are not generating a sample with statistics equal to the specified values. See Section 6–5 of this manual/workbook for a procedure that can be used to generate a sample with a specified mean and standard deviation.

6-2 Finding Areas and Values with a Normal Distribution

The textbook describes methods for working with standard and nonstandard normal distributions. (A **standard normal distribution** has a mean of 0 and a standard deviation of 1.) Table A-2 in Appendix B of the textbook lists a wide variety of different z scores along with their corresponding areas. Minitab can also be used to find probabilities associated with the normal distribution. Minitab is much more flexible than the table, and it includes many more values. In the following procedure, note that like Table A-2 in the textbook, Minitab areas are to the *left* of the related x value. Here is the procedure for finding areas similar to those in the body of Table A-2 from the textbook.

Minitab Procedure for Finding Areas and Values with a Normal Distribution
1.   Select **Calc** from the main menu at the top of the screen.
2.   Select **Probability Distributions** from the subdirectory.
3.   Select **Normal**.
4.   You will now get a dialog box like the one shown below.

When making choices and entries in the dialog box, consider the following:

- **Probability density** is the height of the normal distribution curve, so we will rarely use this option.
- **Cumulative probability:** Use this option to find an area or probability. Note that you will get the area under the curve to the *left* of the *x* value(s) entered in column C1. (You must have already entered the *x* value(s) in column C1.)
- **Inverse cumulative probability:** Use this option to find a value corresponding to a given area. The resulting *x* value(s) separates an area to the *left* that is specified in column C1. (You must have already entered the left areas in column C1.)

The input column is the column containing values (either areas or *x* values) that you have already entered, and the output column is the column that will list the results (either *x* values or areas).

5.   Click **OK** and the results will be listed in the output column.

Examples: Assume that IQ scores are normally distributed with a mean of 100 and a standard deviation of 15.

- To find the area to the left of 115, enter 115 in column C1, then select **Calc**, **Probability Distributions**, then **Normal**. In the dialog box, select the option of **Cumulative Probability** (because you want an *area*). The result will be 0.843415, which is the area to the *left* of 115.

- To find the 90th percentile, enter 0.9 in column C1, then select **Calc**, **Probability Distributions**, then **Normal**. In the dialog box, select the option of **Inverse Cumulative Probability** (because you want a *value* of IQ score). The result will be 119.223, which is the IQ score separating an area of 0.9 to its *left*.

6-3 The Central Limit Theorem

The textbook introduces the central limit theorem, which is then used in subsequent chapters when the important topics of estimating parameters and hypothesis testing are discussed. See the statement of the central limit theorem in the textbook.

In the study of methods of statistical analysis, it is extremely helpful to have a clear understanding of the statement of the central limit theorem. It is helpful to use Minitab in conducting an experiment that illustrates the central limit theorem. For example, let's use Minitab to generate five columns of data randomly selected with Minitab's **Integer** distribution, which generates integers so that they are all equally likely. After generating 5000 values (each between 0 and 9) in columns C1, C2, C3, C4, and C5, we will stack the 25000 values in column

C6. We will also calculate the 5000 sample means of the values in columns C1, C2, C3, C4, and C5 and enter the results in column C7. (In Minitab Release 14, the sample means can be obtained by clicking on **Calc** and selecting **Row Statistics**; select **Mean** and enter C1-C7 for Input variables.) See the histograms shown below. The first histogram shows the distribution of the 25,000 generated integers, but the second histogram shows the distribution of the 5000 sample means. We can clearly see that the sample means have a distribution that is approximately normal.

Distribution of 25,000 randomly selected integers

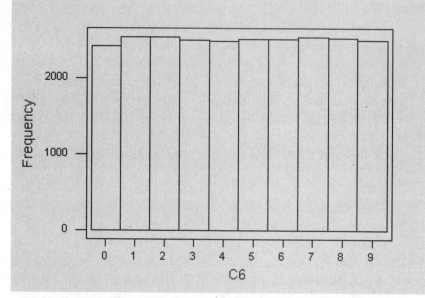

Distribution of 5000 sample means

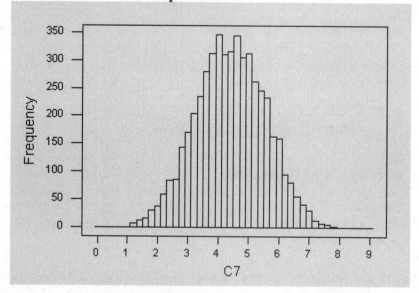

6-4 Assessing Normality

Textbooks in the Triola statistics series discuss criteria for determining whether sample data appear to come from a population having a normal distribution. (This section is not included in *Essentials of Statistics*.) These criteria are listed:

1. **Histogram:** Construct a Histogram. Reject normality if the histogram departs dramatically from a bell shape. Minitab can generate a histogram.

2. **Outliers:** Identify outliers. Reject normality if there is more than one outlier present. (Just one outlier could be an error or the result of chance variation, but be careful, because even a single outlier can have a dramatic effect on results.) Using Minitab, we can sort the data and easily identify any values that are far away from the majority of all other values.

3. **Normal Quantile Plot:** If the histogram is basically symmetric and there is at most one outlier, construct a *normal quantile plot*. Examine the normal quantile plot and reject normality if the points do not lie close to a straight line, or if the points exhibit some systematic pattern that is not a straight-line pattern. Minitab can generate a normal probability plot, which can be interpreted the same way as the normal quantile plot described in the textbook. (Select **Stat**, then **Basic Statistics**, then **Normality Test**.)

Normal Probability Plot To obtain a normal probability plot, click on the main menu item of **Stat**, then select **Basic Statistics**, then **Normality Test**. You will get a dialog box like the one shown below. The entries in the window at the left are present because the worksheet BOSTRAIN was opened so that columns C1 through C7 list sample data. Click on the desired column so that its label appears in the "Variable" box. Leave the reference probability box empty and click **OK**. The resulting normal quantile plot will be as shown on the next page.

Using the normal probability plot shown below, we see that the plot does not yield a pattern of points that reasonably approximates a straight line, so we conclude that rainfall amounts in Boston on Sunday do not appear to be normally distributed.

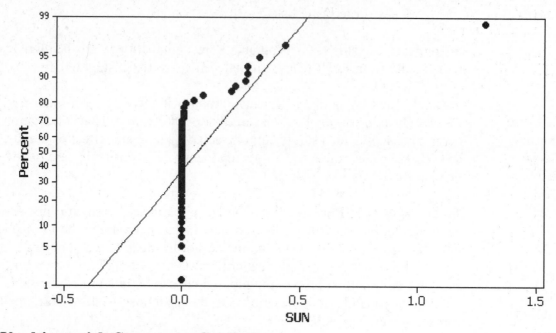

6-5 Working with Summary Statistics

Later chapters in the textbook and later chapters in this manual/workbook develop procedures for using sample data to make inferences about a population. Chapter 7 introduces methods for using sample data to estimate values of population parameters, such as the mean μ or standard deviation σ. Here is a typical problem:

> **Shoveling Heart Rates** In a study of cardiac demands of snow shoveling, 10 subjects used manual shoveling to clear tracts of snow, and their maximum heart rates (beats per minute) were recorded with these results (based on data from "Cardiac Demands of Heavy Snow Shoveling," by Franklin et al., Journal of the American Medical Association, Vol. 273, No. 11):
> $$n = 10$$
> $$\overline{x} = 175$$
> $$s = 15$$
> Use the sample data to estimate the value of the population mean.

The textbook provides procedures for developing estimates of the type suggested by the above statement. In this manual/workbook, we describe methods for using Minitab to solve such problems. Previous versions of Minitab generally provide programs for working with the *original list* of raw data only, but Minitab Release 14 now allows you to use *summary statistics*. This is a major improvement in Minitab Release 14.

Minitab Release 14 now provides programs for working with summary statistics!

If using Minitab Release 13 or earlier, there is often a need to circumvent the obstacle of deal-ing with programs that require the original list of raw data. Here is a procedure that could be applied by those using Minitab Release 13 or earlier.

If Minitab requires an original list of sample values, but we know only the summary statistics, use Minitab to artificially create another sample of the same size with the exact same sample statistics.

The specific method we will use is based on these principles: (1) When you add the same constant to every value in a data set, the mean changes by that same constant, but the standard deviation does not change. (2) When you multiply every value in a data set by the same constant, the mean and standard deviation are both multiplied by that same constant. We will generate a sample with the required number of values, then we will transform it so that it has the desired mean and standard deviation.

Procedure for Creating a Sample with a Specific Mean and Standard Deviation

The following procedure applies to the creation of a sample consisting of 10 values with a mean of 175 and a standard deviation of 15, but any other sample size, mean, and standard deviation could be used.

1.   Select **Calc**, **Random Data**, **Normal**, and proceed to fill out the dialog box as shown below, where the sample size of 10 can be changed as desired. (Leave the default values of 0 and 1 for the mean and standard deviation.) Click **OK**.

2. The sample values in column C1 will not have the desired mean and standard deviation, so transform them (using commands in the session window or using **Calc, Calculator**) to replace column C1 with the result of the following, where s and $\bar{x}$ are replaced with the specific values desired.

Session command window: **LET C1 = ((C1-MEAN(C1))/STDEV(C1))*s + $\bar{x}$**

Calc/Calculator: **((C1-MEAN(C1))/STDEV(C1))*s + $\bar{x}$**

For example, the following dialog box will transform the values in column C1 that were created from the preceding box. The result will be a list of 10 values in column C1, and those values will have a mean of 175 and a standard deviation of 15. When Minitab requires the sample data, we could use this list even though we don't really know the original sample values.

Using the above procedure, we can create a list of 10 sample values having a mean of 175 and a standard deviation of 15, and that list can be treated as if it consisted of the original sample values. Whereas earlier versions of Minitab might not be able to work with the summary statistics of $n = 10$, $\bar{x} = 175$, and $s = 15$, it can be used with the artificially–created list of 10 values having a mean of 175 and a standard deviation of 15.

CHAPTER 6 EXPERIMENTS: Normal Distributions

6-1. *Finding Probabilities for a Normal Distribution*   Use Minitab's **Normal** probability distribution module to find the indicated probabilities. First select **Calc** from the main menu, then select **Probability Distributions**, then **Normal**. Probabilities can be found by using the **Cumulative Probability** option. Remember that like Table A-2 in the textbook, Minitab's probabilities correspond to cumulative areas from the left.

    a.   Given a population with a normal distribution, a mean of 0, and a standard deviation of 1, find the probability of a value less than 1.50._____

    b.   Given a population with a normal distribution, a mean of 100, and a standard deviation of 15, find the probability of a value less than 120._____

    c.   Given a population with a normal distribution, a mean of 75, and a standard deviation of 10, find the probability of a value greater than 80._____

    d.   Given a population with a normal distribution, a mean of 200, and a standard deviation of 20, find the probability of a value between 180 and 210._____

    e.   Given a population with a normal distribution, a mean of 200, and a standard deviation of 20, find the probability of a value between 205 and 223._____

6-2. *Finding Values for a Normal Distribution*   Use Minitab's **Normal** module to find the indicated vales. First select **Calc** from the main menu, then select **Probability Distributions**, then **Normal**. Values can be found by using the **Inverse cumulative probability** option. Remember that like Table A-2 in the textbook, Minitab's probabilities correspond to cumulative areas from the left.

    a.   Given a population with a normal distribution, a mean of 0, and a standard deviation of 1, what value has an area of 0.1234 to its left?_____

    b.   Given a population with a normal distribution, a mean of 100, and a standard deviation of 15, what value has an area of 0.1234 to its right?_____

    c.   Given a population with a normal distribution, a mean of 75, and a standard deviation of 10, what value has an area of 0.9 to its left?_____

    d.   Given a population with a normal distribution, a mean of 200, and a standard deviation of 20, what value has an area of 0.9 to its right?_____

    e.   Given a population with a normal distribution, a mean of 200, and a standard deviation of 20, what value has an area of 0.005 to its left?_____

6-3.   ***Central Limit Theorem*** In this experiment we will illustrate the central limit theorem. See the illustration in Section 6-3 of this manual/workbook.

    a.   Use Minitab to create columns C1, C2, C3, C4, and C5 so that each column contains 250 values that are randomly generated integers between 0 and 9.

    b.   Use **Data/Stack** (or **Manip/Stack** in Minitab Release 13 or earlier) to stack all of the 1250 values in column C6. Obtain a printed copy of the histogram for the sample data in column C6. Also find the mean and standard deviation and enter those values here.
Mean:_____   Standard deviation:_____

    c.   Now create a column C7 consisting of the 250 sample means. The first entry of column C7 is the mean of the first entries in columns C1 through C5, the second entry of C7 is the mean of the second entries in columns C1 through C5, and so on. If using Minitab Release 14, click Calc, select **Row** Statistics, select Mean, and enter C1-C5 for the input variables. Enter C7 as the column for storing the results. If using Minitab Release 13 or earlier, Column C7 can be created with either of the following two methods:
        i.   In the session window, enter the command
LET C7=(C1+C2+C3+C4+C5)/5.
        ii.   Click on **Calc**, then **Calculator**, then use this expression:
(C1+C2+C3+C4+C5)/5.
Obtain a printed copy of the histogram for the sample data in column C7. Also find the mean and standard deviation and enter those values here.
Mean:_____   Standard deviation:_____

    d.   How do these results illustrate the central limit theorem?

_____

_____

6-4.   ***Central Limit Theorem*** In this experiment we will illustrate the central limit theorem. See the illustration in Section 6-3 of this manual/workbook.

    a.   Use Minitab to create columns C1 through C20 in such a way that each column contains 5000 values that are randomly generated from Minitab's **Uniform** distribution with a minimum of 0 and a maximum of 5.

    b.   Stack all of the sample values in column C21. [See part (b) of the preceding exercise.] Obtain a printed copy of the histogram for the sample data in column C21. Also find the mean and standard deviation and enter those values here.
Mean:_____   Standard deviation:_____

    c.   Now create a column C22 consisting of the 5000 sample means. The first entry of column C22 is the mean of the first entries in columns C1 through C20, the second entry of C22 is the mean of the second entries in columns C1 through C20, and so on. [See part (c) of the preceding exercise.]
Obtain a printed copy of the histogram for the sample data in column C22. Also

find the mean and standard deviation and enter those values here.

Mean:_____   Standard deviation:_____

d.   How do these results illustrate the central limit theorem?

_____

_____

6-5.   *Identifying Significance* People generally believe that the mean body temperature is 98.6°F. The body temperature data set in Appendix B of the textbook includes a sample of 106 body temperatures with these properties: The distribution is approximately normal, the sample mean is 98.20°F, and the standard deviation is 0.62°F. We want to determine whether these sample results differ from 98.6°F by a *significant* amount. One way to make that determination is to study the behavior of samples drawn from a population with a mean of 98.6.

a.   Use Minitab to generate 106 values from a normally distributed population with a mean of 98.6 and a standard deviation of 0.62. Use **Stat/Basic Statistics/ Display Descriptive Statistics** to find the mean of the generated sample. Record that mean here:____

b.   Repeat part a nine more times and record the 10 sample means here:

_____

c.   By examining the 10 sample means in part b, we can get a sense for how much sample means vary for a normally distributed population with a mean of 98.6 and a standard deviation of 0.62. After examining those 10 sample means, what do you conclude about the likelihood of getting a sample mean of 98.20? Is 98.20 a sample mean that could easily occur by chance, or is it significantly different from the likely sample means that we expect from a population with a mean of 98.6?

_____

_____

d.   Given that researchers did obtain a sample of 106 temperatures with a mean of 98.20°F, what does their result suggest about the common belief that the population mean is 98.6°F?

_____

_____

6-6.   ***Identifying Significance***   This experiment involves one of the data sets in Appendix B of the textbook: "Weights and Volumes of Cola."

    a.   Open the worksheet COLA and find the mean and standard deviation of the sample consisting of the volumes of cola in cans of regular Coke. Enter the results here.   Sample mean:_____   Standard deviation:_____

    b.   Generate 10 different samples, where each sample has 36 values randomly selected from a normally distributed population with a mean of 12 oz and a standard deviation of 0.115 oz (based on the claimed volume printed on the cans and the data in Appendix B). For each sample, record the sample mean and enter it here.

_____

    c.   By examining the 10 sample means in part b, we can get a sense for how much sample means vary for a normally distributed population with a mean of 12 and a standard deviation of 0.115. After examining those 10 sample means, what do you conclude about the likelihood of getting a sample mean like the one found for the sample volumes in Appendix B? Is the mean for the sample a value that could easily occur by chance, or is it significantly different from the likely sample means that we expect from a population with a mean of 12?

_____

    d.   Consider the sample mean found from the volumes of regular Coke listed in Appendix B from the textbook. Does it suggest that the population mean of 12 oz (as printed on the label) is not correct?

_____

6-7.   ***Assessing Normality***   Refer to the indicated Minitab data file. In each case, print a histogram, print a normal probability plot, and identify any outliers. Based on the results, determine whether the sample data appear to come from a normally distributed population.

    a.   FRI from the worksheet BOSTRAIN (amounts of rainfall in Boston on Fridays for a recent year)

            Outliers:_____   Normal Distribution?_____

    b.   CANS111 from the worksheet CANS (axial loads of aluminum cans that are 0.0111 in. thick)

            Outliers:_____   Normal Distribution?_____

(continued)

c. MCGWIRE from the worksheet HOMERUNS (distances of homeruns hit by Mark McGwire in 1998)

Outliers:_____ Normal Distribution?_____

d. LENGTH values from the worksheet BEARS (lengths of bears anesthetized and measured)

Outliers:_____ Normal Distribution?_____

e. TOTAL from the worksheet GARBAGE (total weights of garbage discarded by households)

Outliers:_____ Normal Distribution?_____

6-8. *Dealing with Summary Statistics*

a. Use Minitab to create a sample of 10 IQ scores with a mean of exactly 100 and a standard deviation of exactly 15. List the values here.

_____

b. Assume that a sample of 10 IQ scores is known to have a mean of 100 and a standard deviation of 15, but the original IQ scores are not known. After creating a list of 10 values with a mean of 100 and a standard deviation of 15 in part (a), what important characteristics of the original data set may have been changed?

_____

_____

6-9. *Dealing with Summary Statistics* Use Minitab to create a sample of 20 values with a mean of exactly 98.20 and a standard deviation of exactly 0.62. List the values here.

_____

6-10. *Dealing with Summary Statistics* Use Minitab to create a sample of 30 values with a mean of exactly 1.23 and a standard deviation of exactly 0.45. List the values here.

_____

_____

_____

7

Confidence Intervals and Sample Sizes

7-1 Working with Summary Statistics

The textbook describes methods for constructing confidence interval estimates of a population proportion, mean, or standard deviation. Minitab Release 13 and earlier require that you have a list of the original sample data. However, Minitab Release 14 now allows you to use the summary statistics (n, $\bar{x}$, s). The following table shows confidence interval requirements for Minitab Release 13 and 14. If Minitab requires an original list of sample values but only the sample statistics are known, see Section 6-5 of this manual/workbook for a way to work around the requirement of original data.

	Minitab Release 13	Minitab Release 14
Confidence interval for p	Use summary statistics or original sample values.	Use summary statistics or original sample values
Confidence interval for μ	Use original sample values.	Use summary statistics or use original sample values.
Confidence interval for σ	Use original sample values.	Use original sample values.

7-2 Confidence Intervals for Estimating p

The textbook introduces confidence intervals as a way to estimate a population parameter. A **confidence interval** (or **interval estimate**) is a range (or an interval) of values used to estimate the true value of a population parameter.

Important note about methods: The textbook gives a method for constructing a confidence interval estimate of a population proportion p. The textbook procedure is based on using a normal distribution as an approximation to a binomial distribution. However, *Minitab uses an* exact *calculation instead of a normal approximation.* (Minitab provides an option for using the normal approximation method, but it is better to use the exact procedure.)

Finding x: Minitab can use the sample size n and the number of successes x to construct a confidence interval estimate of a population proportion. In some cases, the values of x and n are both known, but in other cases the given information may consist of n and a sample percentage. For example, suppose we know that among 829 people surveyed, 51% said that they are against using cameras to identify and ticket drivers who run red lights. Based on that information, we know that $n = 829$ and $\hat{p} = 0.51$. Because $\hat{p} = x/n$, it follows that $x = \hat{p}n$, so the number of successes can be found by multiplying the sample proportion p and the sample size n. Given that $n = 829$ and $\hat{p} = 0.51$, we find $x = (0.51)(829) = 422.79$, which we round to the whole number 423. It's actually quite simple: 51% of 829 is $0.51 \times 829 = 423$ (rounded to a whole number).

> **To find the number of successes x from the sample proportion and sample size:**
> **Calculate $x = \hat{p}\, n$ and round the result to the nearest whole number.**

After having determined the value of the sample size n and the number of successes x, we can proceed to use Minitab as follows.

Minitab Procedure for Finding Confidence Intervals for p

1. Select **Stat** from the main menu.

2. Select **Basic Statisticss**.

3. Select **1 Proportion**.

4. You will get a dialog box like the one shown below. The entries correspond to $n = 829$ and $x = 423$.

5. Click on **Options** to change the confidence level from 95% to any other value. (The **Options** button can also be used to check the box for "Test and interval based on normal distribution." You could click on that box to use the same procedure given in the textbook, but it is better to use the *exact* procedure, so leave that box unchecked.)

After clicking **OK**, the Minitab results are as shown below.

```
Test and CI for One Proportion

Test of p = 0.5 vs p not = 0.5

                                                         Exact
Sample      X       N   Sample p            95.0% CI    P-Value
1         423     829   0.510253   (0.475636, 0.544797)   0.578
```

From the above Minitab results, we see that the 95% confidence interval estimate of p is displayed as (0.475636, 0.544797). Rounding the confidence interval limits to three decimal places, we get (0.476, 0.545). This confidence interval can also be expressed in the format of $0.476 < p < 0.545$.

7-3 Confidence Intervals for Estimating μ

The textbook introduce confidence intervals used for estimating the value of a population mean μ. The textbook stresses the importance of selecting the correct distribution (normal or t). The textbook provides a summary of the criteria, and the table below is included.

Choosing between z and t

Method	Conditions
Use normal (z) distribution.	**σ known and normally distributed population** *or* **σ known and $n > 30$**
Use t distribution.	**σ not known and normally distributed population** *or* **σ not known and $n > 30$**
Use a nonparametric method or bootstrapping.	**Population is not normally distributed and $n \leq 30$**

The textbook also notes that in choosing between the normal or t distributions, a major criterion for choosing between the normal (z) or Student t distributions is whether the population standard deviation σ is known. In reality, it is very rare to know σ, so the t distribution is typically used instead of the normal distribution.

Here is the procedure for using Minitab to construct confidence interval estimates of a population mean μ.

Minitab Procedure for Finding Confidence Intervals for μ

1. With Minitab Release 14, use either the original sample values or the summary statistics of n, $\bar{x}$, and s. (If using Minitab Release 13 or earlier and the original sample data values are not known, *generate* a list of sample values with the same sample size, the same mean, and the same standard deviation. Do this by using the method described in Section 6-5 of this manual/workbook.)

2. Select **Stat** from the main menu.

3. Select **Basic Statistics** from the subdirectory.

4. Select either **1-Sample** z or **1-Sample t** by using the criteria summarized in the preceding table.

5.   You will now see a dialog box, such as the one shown below. If using Minitab Release 14 or later, the dialog box will include entries for the summarized data consisting of the sample size, sample mean, and sample standard deviation, as shown in the display below. (If you select 1-Sample z in Step 4, there will also be a box for entering "Sigma," the population standard deviation σ.)

   Note: If using an earlier version of Minitab, the "Summarized data" option is not available, and you must use either the original list of sample values or you must create a list of sample values with the same size, mean, and standard deviation as the original sample (as described in Section 6-5 of this manual/workbook).

6.   In the "Variables" box, enter the column containing the list of sample data. If using Minitab Release 14 and the summary statistics are known, click **Summarized data** and proceed to enter the sample size, sample mean, and sample standard deviation, as shown in the display below. Click on the **Options** box to enter the confidence level. (The Minitab default is a confidence level of 95%.) Click **OK** when done.

1-Sample t (Test and Confidence Interval) ☒

○ **Samples in columns:**

⊙ **Summarized data**

Sample size:	190
Mean:	2700
Standard deviation:	645

☐ **Perform hypothesis test**

Hypothesized mean:

Select		Graphs...	Options...
Help		OK	Cancel

As an illustration, consider this example:

> **Confidence Interval for Birth Weights** In a study of the effects of prenatal cocaine use on infants, the following sample data were obtained for weights at birth: $n = 190$, $\bar{x} = 2700$ g, $s = 645$ g (based on data from "Cognitive Outcomes of Preschool Children With Prenatal Cocaine Exposure" by Singer et al, *Journal of the American Medical Association,* Vol. 291, No. 20). The design of the study justifies the assumption that the sample can be treated as a simple random sample. Use the sample data to construct a 95% confidence interval estimate of μ, the mean birth weight of all infants born to mothers who used cocaine.

The preceding Minitab dialog box shows the required entries. After clicking **OK**, the Minitab results will be as shown below.

One-Sample T

```
  N    Mean    StDev   SE Mean       95% CI
190  2700.00  645.00    46.79   (2607.70, 2792.30)
```

The above Minitab display includes the 95% confidence interval limits of 2607.70 and 2792.30. We can round those limits (using the same number of decimal places as in the sample mean and standard deviation) and express the confidence interval in this format:

$$2608 \text{ g} < \mu < 2792 \text{ g}$$

7-4 Confidence Intervals for Estimating σ

Here is the Minitab procedure for constructing confidence interval estimates of the population standard deviation σ. (If you want a confidence interval for the variance σ^2, use the following procedure and square the resulting confidence interval limits.) Note that this procedure requires a list of the original sample data. If only the summary statistics are known, use the procedure in Section 6-5 of this manual/workbook to generate a list of sample values.

Minitab Procedure for Confidence Intervals for Estimating σ

1. If the original sample values are known, enter them in a Minitab column.

    If the original sample data values are not known, *generate* a list of sample values with the same sample size, the same mean, and the same standard deviation. Do this by using the method described in Section 5-5 of this manual/workbook.

2. Select **Stat** from the main menu.

3.   Select **Basic Statistics** from the subdirectory.

4.   If using Minitab Release 14, select **Graphical** Summary. (If using earlier releases of Minitab, select **Display Descriptive Statistics**.)

5.   You will get a dialog box that includes a "Variables" box. In that box, enter the column containing the data. Also enter the confidence level. Click **OK**. (If using Minitab Release 13 or earlier, click on the Graph button, select Graphical Summary, and enter the confidence level.)

Consider this example:

> **Body Temperatures**   Appendix B in the textbook lists 106 body temperatures (at 12:00 AM on day 2) obtained by University of Maryland researchers. Use the following characteristics of the data set to construct a 95% confidence interval estimate of σ, the standard deviation of the body temperatures of the whole population:
> **a.** A histogram of the sample data suggests a normally distributed population.
> **b.** The sample mean is 98.20°F.
> **c.** The sample standard deviation is $s = 0.62$°F.
> **d.** The sample size is $n = 106$.
> **e.** There are no outliers.

If you use the Minitab procedure with the 106 body temperatures (at 12:00 AM on day 2) listed in the "Body Temperatures" data set from Appendix B of the textbook, the result will be as shown on the next page. See the lower right corner of the Minitab display. The 95% confidence interval limits for σ are displayed as 0.549 and 0.720. We can round these confidence interval limits to get a 95% confidence interval: $0.55 < \sigma < 0.72$. (To round, use one more decimal place than the single decimal place used in the original list of data values.) Also, squaring the original confidence interval limits and rounding the squared values results in this confidence interval limit for the variance: $0.30 < \sigma^2 < 0.52$.

Summary for C1

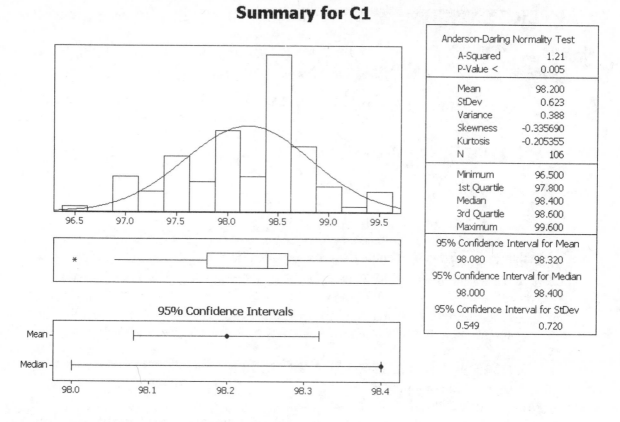

Anderson-Darling Normality Test	
A-Squared	1.21
P-Value <	0.005

Mean	98.200
StDev	0.623
Variance	0.388
Skewness	-0.335690
Kurtosis	-0.205355
N	106

Minimum	96.500
1st Quartile	97.800
Median	98.400
3rd Quartile	98.600
Maximum	99.600

95% Confidence Interval for Mean	
98.080	98.320

95% Confidence Interval for Median	
98.000	98.400

95% Confidence Interval for StDev	
0.549	0.720

7-5 Determining Sample Sizes

The textbook discuss procedures for determining sample sizes necessary to estimate a population proportion p, or mean μ, or standard deviation σ. Minitab does not have the ability to calculate sample sizes directly, but determinations of sample size use simple formulas, so it is easy to use a calculator with the procedures described in the textbook. Minitab could be used instead of a calculator. Consider this formula for determining the sample size required to estimate a population mean:

$$n = \left[\frac{z_{\alpha/2}\,\sigma}{E} \right]^2$$

Suppose that we want to find the sample size needed to estimate the mean IQ score of statistics professors, and we want 95% confidence that the error in the sample mean is no more than 2 IQ points. We further assume that $\sigma = 15$. The required value of z can be found by using the procedure described in Section 6-2 of this manual/workbook. For example, to find the z score corresponding to 95% in two tails, select **Inverse cumulative probability** and enter a cumulative left area of 0.975 to get the z score of 1.95996. Then enter the Minitab command LET C2=(1.95996*15/2)**2 to complete the calculation for the sample size. The result will be the sample size of 216.082, which we round up to 217. The determination of sample size for estimating a population proportion p can be handled the same way. The determination of sample size for estimating σ or σ^2 can be done by using the table given in the textbook in the section "Estimating a Population Variance."

CHAPTER 7 EXPERIMENTS: Confidence Intervals and Sample Sizes

In Experiments 7–1 through 7–4, use Minitab with the sample data and confidence level to construct the confidence interval estimate of the population proportion p.

7–1. $n = 400, x = 300$, 95% confidence   _____

7–2. $n = 1200, x = 200$, 99% confidence   _____

7–3. $n = 1655, x = 176$, 98% confidence   _____

7–4. $n = 2001, x = 1776$, 90% confidence   _____

7–5.   ***Internet Shopping*** In a Gallup poll, 1025 randomly selected adults were surveyed and 29% of them said that they used the Internet for shopping at least a few times a year.

   a.   Find the point estimate of the percentage of adults who use the Internet for shopping. _____

   b.   Find a 99% confidence interval estimate of the percentage of adults who use the Internet for shopping. _____

   c.   If a traditional retail store wants to estimate the percentage of adult Internet shoppers in order to determine the maximum impact of Internet shoppers on its sales, what percentage of Internet shoppers should be used? _____

7–6.   ***Death Penalty Survey*** In a Gallup poll, 491 randomly selected adults were asked whether they are in favor of the death penalty for a person convicted of murder, and 65% of them said that they were in favor.

   a.   Find the point estimate of the percentage of adults who are in favor of this death penalty. _____

   b.   Find a 95% confidence interval estimate of the percentage of adults who are in favor of this death penalty. _____

   c.   Can we safely conclude that the majority of adults are in favor of this death penalty? Explain.

   _____

   _____

7–7. ***Mendelian Genetics*** When Mendel conducted his famous genetics experiments with peas, one sample of offspring consisted of 428 green peas and 152 yellow peas.

    a.   Use Minitab to find the following confidence interval estimates of the percentage of yellow peas.

        99.5% confidence interval: _____
        99% confidence interval: _____
        98% confidence interval: _____
        95% confidence interval: _____
        90% confidence interval: _____

    b.   After examining the pattern of the above confidence intervals, complete the following statement. "As the degree of confidence decreases, the confidence interval limits _____."

    c.   In your own words, explain why the preceding completed statement makes sense. That is, why should the confidence intervals behave as you have described?

7–8. ***Misleading Survey Responses*** In a survey of 1002 people, 701 said that they voted in a recent presidential election (based on data from ICR Research Group). Voting records show that 61% of eligible voters actually did vote. Find a 99% confidence interval estimate of the proportion of people who say that they voted. _____

7–9. ***Estimating Car Pollution*** In a sample of seven cars, each car was tested for nitrogen-oxide emissions (in grams per mile) and the following results were obtained: 0.06, 0.11, 0.16, 0.15, 0.14, 0.08, 0.15 (based on data from the Environmental Protection Agency). Assuming that this sample is representative of the cars in use, construct a 98% confidence interval estimate of the mean amount of nitrogen-oxide emissions for all cars.

_____

7–10. ***Monitoring Lead in Air*** Listed below are measured amounts of lead (in micrograms per cubic meter or $\mu g/m^3$) in the air. The Environmental Protection Agency has established an air quality standard for lead: 1.5 $\mu g/m^3$. The measurements shown below were recorded at Building 5 of the World Trade Center site on different days immediately following the destruction caused by the terrorist attacks of September 11, 2001. After the collapse of the two World Trade buildings, there was considerable concern about the quality of the air. Use the given values to construct a 95% confidence interval estimate of the mean amount of lead in the air. Is there anything about this data set suggesting that the confidence interval might not be very good? Explain.

    5.40   1.10   0.42   0.73   0.48   1.10

7–11. *Weights of Discarded Garbage*   Refer to the Garbage data set in Appendix B of the textbook. The Minitab worksheet name is GARBAGE. Find a 95% confidence interval estimate of the mean total weight of discarded garbage. _____

7–12. *Axial Loads of Aluminum Cans*   Refer to the axial loads of aluminum cans in the data set from Appendix B. The Minitab worksheet name is CANS.

    a.   Find a 95% confidence interval estimate of the population mean of axial loads of cans that are 0.0109 in. thick. _____

    b.   Find a 95% confidence interval estimate of the population mean of axial loads of cans that are 0.0111 in. thick. _____

    c.   Compare the results from parts (a) and (b).

          _____

7–13. *Weights of Bears* The health of the bear population in Yellowstone National Park is monitored by periodic measurements taken from anesthetized bears. A sample of 54 bears has weights listed in the bears data set in Appendix B in the textbook. The Minitab worksheet is BEARS.

    a.   Assuming that σ is known to be 121.8 lb, find a 99% confidence interval estimate of the mean of the population of all such bear weights. _____

    b.   Find a 99% confidence interval estimate of the mean of the population of all such bear weights, assuming that the population standard deviation σ is not known.

          _____

    c.   Compare the results from parts (a) and (b).

          _____

7–14. *Pulse Rates of Men and Women*   Refer to Data Set 1 in Appendix B of the textbook. The Minitab worksheet name for men is MHEALTH and the worksheet name for women is FHEALTH.

    a.   Find a 95% confidence interval estimate of the mean pulse rate of men.

          _____

    b.   Find a 95% confidence interval estimate of the mean pulse rate of women.

          _____

    c.   Compare the confidence intervals for men and women. What do you conclude?

          _____

7–15. **Body Mass Index** Refer to Data Set 1 in Appendix B and use the sample data.

    a.   Construct a 99% confidence interval estimate of the standard deviation of body mass indexes for men.

_____

    b.   Construct a 99% confidence interval estimate of the standard deviation of body mass indexes for women

_____

    c.   Compare and interpret the results.

_____

_____

7–16. **Using Summary Statistics** A study was conducted to estimate hospital costs for accident victims who wore seat belts. Twenty randomly selected cases have a distribution that appears to be bell-shaped with a mean of $9004 and a standard deviation of $5629 (based on data from the U.S. Department of Transportation). Construct the 99% confidence interval for the mean of all such costs. _____

7–17. **Using Summary Statistics**   Because cardiac deaths appear to increase after heavy snowfalls, an experiment was designed to compare cardiac demands of snow shoveling to those of using an electric snow thrower. Ten subjects cleared tracts of snow using both methods, and their maximum heart rates (beats per minute) were recorded during both activities. The following results were obtained (based on data from "Cardiac Demands of Heavy Snow Shoveling," by Franklin et al., Journal of the American Medical Association, Vol. 273, No. 11):

        Manual Snow Shoveling Maximum Heart Rates: $n = 10$, $\bar{x} = 175$, $s = 15$
        Electric Snow Thrower Maximum Heart Rates: $n = 10$, $\bar{x} = 124$, $s = 18$

    a.   Find the 95% confidence interval estimate of the population mean for those people who shovel snow manually._____

    b.   Find the 95% confidence interval estimate of the population mean for those who people who use the electric snow thrower. _____

    c.   Compare the results from parts (a) and (b).

_____

7-18.   ***Confidence Interval for Estimating a Mean***   A data set in Appendix B of the textbook lists sample weights of the cola in cans of regular Coke. Those weights are listed below, but they are also stored as a column in the Minitab worksheet COLA, so it is not necessary to manually enter these values. Instead, open the worksheet COLA.

Weights (in pounds) of a sample of cans of regular Coke

0.8192	0.8150	0.8163	0.8211	0.8181	0.8247
0.8062	0.8128	0.8172	0.8110	0.8251	0.8264
0.7901	0.8244	0.8073	0.8079	0.8044	0.8170
0.8161	0.8194	0.8189	0.8194	0.8176	0.8284
0.8165	0.8143	0.8229	0.8150	0.8152	0.8244
0.8207	0.8152	0.8126	0.8295	0.8161	0.8192

a.   Use Minitab to find the following confidence interval estimates of the population mean.

99.5% confidence interval: _____

99% confidence interval: _____

98% confidence interval: _____

95% confidence interval: _____

90% confidence interval: _____

b.   Change the first weight from 0.8192 lb to 8192 lb (a common error in data entry) and find the indicated confidence intervals for the population mean.

99.5% confidence interval: _____

99% confidence interval: _____

98% confidence interval: _____

95% confidence interval: _____

90% confidence interval: _____

c.   By comparing these results from parts (a) and (b), what do you conclude about the effect of an outlier on the values of the confidence interval limits?

_____

7-19. ***Simulated Data*** Minitab is designed to generate random numbers from a variety of different sampling distributions. In this experiment we will generate 500 IQ scores, then we will construct a confidence interval based on the sample results. IQ scores have a normal distribution with a mean of 100 and a standard deviation of 15. First generate the 500 sample values as follows.

1. Click on **Calc,** then select **Random Data**, then **Normal**.

2. In the dialog box, enter a sample size of 500, a mean of 100, and a standard deviation of 15. Specify column C1 as the location for storing the results. Click **OK**.

3. Use **Stat/Basic Statistics/Display Descriptive Statistics** to find these statistics:

$n = $ _____ $\bar{x} = $ _____ $s = $ _____

Using the generated values, construct a 95% confidence interval estimate of the population mean of all IQ scores. Enter the 95% confidence interval here.

_____

Because of the way that the sample data were generated, we *know* that the population mean is 100. Do the confidence interval limits contain the true mean IQ score of 100?

_____

If this experiment were to be repeated over and over again, how often would we expect the confidence interval limits to contain the true population mean value of 100? Explain how you arrived at your answer.

_____

_____

7-20. ***Simulated Data*** Follow the same steps listed in Experiment 7-19 to randomly generate 500 IQ scores from a population having a normal distribution, a mean of 100, and a standard deviation of 15. Record the sample statistics here.

$n = $ _____ $\bar{x} = $ _____ $s = $ _____

Confidence intervals are typically constructed with confidence levels around 90%, 95%, or 99%. Instead of constructing such a typical confidence interval, use the generated values to construct a 50% confidence interval. Enter the result below.

(continued)

Does the above confidence interval have limits that actually do contain the true population mean, which we know is 100?_____

Repeat the above procedure 9 more times and list the resulting 50% confidence intervals here.

_____ _____ _____ _____ _____

_____ _____ _____ _____

Among the total of the 10 confidence intervals constructed, how many of them actually do contain the true population mean of 100? Is this result consistent with the fact that the level of confidence used is 50%? Explain.

_____

_____

7-21. ***Combining Data Sets*** Refer to the M&M data set in Appendix B and use the entire sample of 100 plain M&M candies to construct a 95% confidence interval for the mean weight of all M&Ms. (*Hint*: It is not necessary to manually enter the 100 weights, because they are already stored in separate Minitab columns in the worksheet M&M. Use **Stack** to combine the different M&M data sets into one big data set. Now construct a 95% confidence interval estimate of the population mean of all IQ scores. Enter the 95% confidence interval here.

_____

7-22. ***Estimating Standard Deviation*** Refer to the sample data used in Experiment 7-18 and use Minitab to construct a 95% confidence interval to estimate the population standard deviation σ. Enter the result here.

_____

Refer to the textbook and identify the requirements for the procedures used to construct confidence intervals for estimating a population standard deviation.

_____

_____

Assuming that the sample is a simple random sample, how can the other requirement be checked? Can Minitab be used to check for normality? If so, do such a check and report the results and conclusion here.

_____

7-23. ***Estimating Standard Deviation***   A container of car antifreeze is supposed to hold 3785 mL of the liquid. Realizing that fluctuations are inevitable, the quality-control manager wants to be quite sure that the standard deviation is less than 30 mL. Otherwise, some containers would overflow while others would not have enough of the coolant. She selects a simple random sample, with the results given here.

3761	3861	3769	3772	3675	3861
3888	3819	3788	3800	3720	3748
3753	3821	3811	3740	3740	3839

Use these sample results to construct the 99% confidence interval for the true value of σ. Does this confidence interval suggest that the fluctuations are at an acceptable level?

7-24. ***Quality Control of Doughnuts***   The Hudson Valley Bakery makes doughnuts that are packaged in boxes with labels stating that there are 12 doughnuts weighing a total of 42 oz. If the variation among the doughnuts is too large, some boxes will be underweight (cheating consumers) and others will be overweight (lowering profit). A consumer would not be happy with a doughnut so small that it can be seen only with an electron microscope, nor would a consumer be happy with a doughnut so large that it resembles a tractor tire. The quality-control supervisor has found that he can stay out of trouble if the doughnuts have a mean of 3.50 oz and a standard deviation of 0.06 oz or less. Twelve doughnuts are randomly selected from the production line and weighed, with the results given here (in ounces).

3.43   3.37   3.58   3.50   3.68   3.61   3.42   3.52   3.66   3.50   3.36   3.42

Construct a 95% confidence interval for σ, then determine whether the quality-control supervisor is in trouble.

Hypothesis Testing

8-1 Working with Summary Statistics

The textbook describes procedures for testing hypotheses about a population proportion, mean, or standard deviation. In some cases, Minitab allows the use of the summary statistics (n, $\bar{x}$, s), but in other cases you must use the original list of sample values. See the table below. If Minitab requires an original list of sample values but only the sample statistics are known, see Section 6-5 of this manual/workbook for a way to work around the requirement of knowing the original data values.

	Minitab Release 13	Minitab Release 14
Hypothesis test for p	Use summary statistics or original sample values.	Use summary statistics or original sample values
Hypothesis test for μ	Use original sample values.	Use summary statistics or original sample values.
Hypothesis test for σ	Use original sample values.	Use original sample values.

8-2 Testing Hypotheses About p

Important note about methods: The textbook gives a method for testing a claim about a population proportion p. The textbook procedure is based on using a normal distribution as an approximation to a binomial distribution. However, *Minitab uses an* exact *calculation instead of a normal approximation.* (Minitab provides an option for using the normal approximation method, but it is better to use the exact procedure.)

> **Minitab uses an *exact* procedure (instead of using a normal approximation) for testing hypothesis about a population proportion p.**

The Minitab procedure for testing claims about a population proportion follows the same basic steps described in Section 7-2 of this manual/workbook. In addition to having a claim to be tested, Minitab also requires a significance level, the sample size n, and the number of successes x.

   Finding x: In Section 7-2 of this manual/workbook, we briefly discussed one particular difficulty that arises when the available information provides the sample size n and the sample proportion $\hat{p}$ instead of the number of successes x. We provided this procedure for determining the number of successes x.

> **To find the number of successes x from the sample proportion and sample size:**
> **Calculate $x = \hat{p}\,n$ and round the result to the nearest whole number.**

Given a claim about a proportion and knowing n and x, we can use Minitab as follows.

Minitab Procedure for Testing Claims about p

1. Select **Stat** from the main menu.

2. Select **Basic Statistics** from the subdirectory.

3. Select **1 Proportion**.

4. You will get a dialog box.

- Click on the option for summarized data.
- In the box for "Number of trials," enter the value of the sample size n.
- In the box for the number of events (or successes), enter the number of successes x.
- Click on the **Options** bar and enter the confidence level. (Enter 95 for a 0.05 significance level.) Also, in the "Test proportion" box, enter the claimed value of the population proportion p. Also select the format of the alternative hypothesis.
- Click **OK**.

For example, let's use the sample data of $n = 880$ and $x = 493$ to test the claim that $p > 0.5$. The dialog boxes and Minitab results are as follows.

Based on the above dialog boxes, the following Minitab results are obtained.

```
Test and CI for One Proportion

Test of p = 0.5 vs p > 0.5
```

					Exact
Sample	X	N	Sample p	95.0% Lower Bound	P-Value
1	493	880	0.560227	0.532017	0.000

We can see from this display that the *P*-value is 0.000, which causes us to reject the null hypothesis and support the claim that $p > 0.5$. The Minitab value of 0.532017 for a "95% Lower Bound" corresponds to this 95% one–sided confidence interval: $p > 0.532017$. That is, based on the sample data, we have 95% confidence that the true population proportion p has a value that is greater than 0.532017.

8–3 Testing Hypotheses About μ

The textbook describes procedures for testing claims about the mean of a single population. When testing claims about the mean of a single population, the textbook explains that there are different procedures, depending on the size of the sample, the nature of the population distribution, and whether the population standard deviation σ is known. The textbook stresses the importance of selecting the correct distribution (normal or *t*). See the following table, which summarizes the criteria for choosing between the normal and *t* distributions.

Choosing between z and t

Method	Conditions
Use normal (z) distribution.	σ known and normally distributed population *or* σ known and $n > 30$
Use t distribution.	σ not known and normally distributed population *or* σ not known and $n > 30$
Use a nonparametric method or bootstrapping.	Population is not normally distributed and $n \leq 30$

Minitab Release 14 allows us to test claims about a population mean μ by using either the list of original sample values or the summary statistics of n, $\overline{x}$, and s. (Minitab Release 13 requires the original list of sample values.)

Minitab Procedure for Testing Claims About μ

1.   With Minitab Release 14, use either the original sample values or the summary statistics of n, $\overline{x}$, and s. (Minitab Release 13 requires a list of the original sample values. If using Minitab Release 13 or earlier, and the original sample data values are not known, *generate* a list of sample values with the same sample size, the same mean, and the same standard deviation. Do this by using the method described in Section 6–5 of this manual/workbook.)

2.   Select **Stat** from the main menu.

3.   Select **Basic Statistics** from the subdirectory.

4.   Select either **1-Sample** z or **1-Sample t** by using the criteria summarized in the above table. (If the methods of this chapter do not apply, you're stuck. You must use other methods, such as bootstrap resampling or nonparametric methods.)

5.   You will now see a dialog box, such as the one shown below. (If using Minitab Release 13 or earlier, the dialog box will not include the entries for the summarized data consisting of the sample size, sample mean, and sample standard deviation.) If you select 1-Sample z in Step 4, there will also be a box for entering "Sigma," the population standard deviation σ.

6.   In the "Samples in columns" box, enter the column containing the list of sample data. If using Minitab Release 14 and the summary statistics are known, click **Summarized data** and proceed to enter the sample size, sample mean, and sample standard deviation.

7.   Click on the box labeled "Perform hypothesis test."

8.   In the "Hypothesized mean" box, enter the claimed value of the population mean.

9.   Click on the **Options** button to get the options dialog box, and make the entries according to the following.

- Enter a confidence level. (Enter 95 for a hypothesis test conducted with a 0.05 significance level.)

- Click on the "Alternative" box to change the default alternative hypothesis. The available choices in the "Alternative" box are
    less than
    not equal
    greater than

- Click **OK** when done.

10.   Click **OK** on the main dialog box.

As an illustration, consider this example:

Body Temperatures A pre–med student in a statistics class is required to do a class project. Intrigued by the Body Temperatures data set in Appendix B, she plans to collect her own sample data to test the claim that the mean body temperature is less than 98.6°F, as is commonly believed. Because of time constraints imposed by other courses and the desire to maintain a social life that goes beyond talking in her sleep, she finds that she has time to collect data from only 12 people. After carefully planning a procedure for obtaining a simple random sample of 12 healthy adults, she measures their body temperatures and obtains the results listed below. Use a 0.05 significance level to test the claim that these body temperatures come from a population with a mean that is less than 98.6°F.

98.0 97.5 98.6 98.8 98.0 98.5 98.6 99.4 98.4 98.7 98.6 97.6

After using Minitab to enter the given sample data in column C1, we apply the preceding Minitab procedure for testing a claim about a population mean ($\mu < 98.6$). The population standard deviation σ is not known and the sample is small, so we first test for normality by generating a histogram. (For the procedure used to generate a histogram, see Section 2–1 of this manual/workbook).

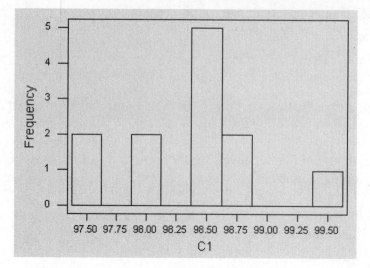

Because the above histogram is not dramatically different from a normal distribution, we assume that the sample data are from a population having a normal distribution. (Remember, the *t* test requirement of a normal distribution is somewhat loose.) We now proceed to conduct the hypothesis test using the *t* distribution. (We have normally distributed data and σ is not known. Given these conditions, the *t* distribution should be used instead of the normal distribution.)

The following dialog boxes correspond to the example we are considering. (The sample data are in column C1 and we are using a 0.05 significance level to test the claim that $\mu < 98.6$.)

We get these Minitab results:

One-Sample T: C1

```
Test of mu = 98.6 vs < 98.6
                                                95%
                                              Upper
Variable    N     Mean    StDev   SE Mean    Bound       T       P
C1         12   98.3917   0.5351   0.1545   98.6691   -1.35   0.102
```

From the above Minitab display, we see that the test statistic is $t = -1.35$. The P-value is 0.102. (Minitab does not provide the critical value, so we use the P-value method of hypothesis testing.) Because the P-value is greater than the significance level of 0.05, we fail to reject the null hypothesis and conclude that there is not sufficient evidence to support the claim that the population mean is less than 98.6. This does not "prove" that the mean is 98.6°F. In fact, μ may well be less than 98.6°F, but the 12 sample values do not provide evidence strong enough to support that claim

8-4 Testing Hypotheses About σ or σ^2

Minitab is not designed to test claims about σ or σ^2, but it is capable of constructing confidence interval estimates of σ or σ^2 (as described in Section 7–4 of this manual/workbook), so we can use the confidence interval approach to hypothesis testing. The confidence interval approach can be summarized as follows:

> Because a confidence interval estimate of a population parameter contains the likely values of that parameter, reject a claim that the population parameter has a value that is not included in the confidence interval.

Caution: See the textbook where these two important requirements are given for hypothesis tests about σ or σ^2:

1.   The samples are simple random samples. (Remember the importance of good sampling methods.)

2.   The sample values come from a population with a *normal distribution*.

The textbook makes the very important point that for tests of claims about standard deviations or variances, the requirement of a normal distribution is very strict. If the population does not have a normal distribution, then inferences about standard deviations or variances can be very misleading. Suggestion: Given sample data, construct a histogram and/or normal probability plot to determine whether the assumption of a normal distribution is reasonable. If the population distribution does appear to be normal and you want to test a claim about the population standard deviation or variance, use the Minitab procedure given below.

Minitab Procedure for Testing Claims about σ or σ^2

This procedure is based on the use of confidence intervals. We can make a direct correspondence between a confidence interval and a hypothesis test only when the test is two-tailed. A one-tailed hypothesis test with significance level α corresponds to a confidence interval with a confidence level of $1 - 2\alpha$. See the following table for common cases. For example to test $\mu < 98.6$ using a 0.05 significance level, construct a 90% confidence interval.

		Confidence Level Used for Hypothesis Test	
		Two-Tailed Test	One-Tailed Test
Significance Level	0.01	99%	98%
for Hypothesis	0.05	95%	90%
Test	0.10	90%	80%

1.   Construct a confidence interval estimate of σ by using the same procedure outlined in Section 7–4 of this manual/workbook. When selecting the confidence level, see the above table. (For a *two*-tailed test with a 0.05 significance level, use a 95% confidence interval; for a *one*-tailed test with a 0.05 significance level, use a 90% confidence interval.)

2.   Because a confidence interval estimate of a population parameter contains the likely values of that parameter, reject a claim that the population parameter has a value that is not included in the confidence interval.

As an illustration, consider this example:

> ***IQ Scores of Statistics Professors*** IQ scores for adults are normally distributed with a mean of 100 and a standard deviation of 15. A simple random sample of 13 statistics professors yields a standard deviation of $s = 7.2$. A psychologist claims that statistics professors have IQ scores with a standard deviation equal to 15, the same standard deviation for the general population. Use a 0.05 level of significance to test the claim that $\sigma = 15$.

From the above statements, we see that we want to test the claim that $\sigma = 15$, and we want to use a 0.05 significance level. Sample data consist of the summary statistics $n = 13$ and $s = 7.2$, so we must first generate a sample consisting of 13 values with a standard deviation of 7.2, which is accomplished using the procedure described in Section 6–5. We then use the procedure outlined in Section 7–4 of this manual/workbook, and we get this 95% confidence interval: $5.1630 < \sigma < 11.8853$. This confidence interval does not contain the claimed value of $\sigma = 15$, so we reject the null hypothesis that $\sigma = 15$. It appears that there is sufficient evidence to warrant rejection of the claim that statistics professors have IQ scores with a standard deviation equal to 15.

8-5 Testing Hypotheses with Simulations

Sections 8-2, 8-3, and 8-4 of this manual/workbook have all focused on using Minitab for hypothesis tests using the *P*-value approach or the confidence interval approach. Another very different approach is to use *simulations*. Let's illustrate the simulation technique with an example.

In Section 8–3 of this manual/workbook, we tested the null hypothesis that $\mu < 98.6$ for the body temperatures of healthy adults. We used sample data consisting of these 12 values:

98.0 97.5 98.6 98.8 98.0 98.5 98.6 99.4 98.4 98.7 98.6 97.6

For these 12 values, $n = 12$, $\bar{x} = 98.39167$, and $s = 0.53506$. The key question is this:

> If the population mean body temperature is really 98.6, then *how likely* is it that we would get a sample mean of 98.39167, given that the population has a normal distribution and the sample size is 12?

If the probability of getting a sample mean such as 98.39167 is very small, then that suggests that the sample results are not the result of chance random fluctuation. If the probability is high, then we can accept random chance as an explanation for the discrepancy between the sample mean of 98.39167 and the assumed mean of 98.6. What we need is some way of determining the likelihood of getting a sample mean such as 98.39167. That is the precise role of *P*-values in the *P*-value approach to hypothesis testing. However, there is another approach. Minitab and many other software packages are capable of generating random results from a variety of different populations. Here is how Minitab can be used: Determine the likelihood of getting a sample mean of 98.39167 by randomly generating different samples from a population that is normally distributed with the claimed mean of 98.6. For the standard deviation, we will use the best available information: the value of $s = 0.53506$ obtained from the sample.

Minitab Procedure for Testing Hypotheses with Simulations

1.   Identify the values of the sample size n, the sample standard deviation s, and the claimed value of the population mean.

2.   Click on **Calc**, then select **Random Data**.

3.   Click on **Normal.**

4.   Generate a sample randomly selected from a population with the claimed mean. When making the required entries in the dialog box, use the *claimed* mean, the sample size n, and the sample standard deviation s.

5.   Continue to generate similar samples until it becomes clear that the given sample mean is or is not likely. (Here is one criterion: The given sample mean is *unlikely* if its value or more extreme values occur 5% of the time or less.) If it is unlikely, reject the claimed mean. If it is likely, fail to reject the claimed mean.

For example, here are 20 results obtained from the random generation of samples of size 12 from a normally distributed population with a mean of 98.6 and a standard deviation of 0.53506:

98.754	**98.374**	**98.332**	98.638	98.513
98.551	98.566	98.760	**98.332**	98.603
98.407	98.640	98.655	98.408	98.802
98.505	98.699	98.609	98.206	98.582

Examining the 20 sample means, we see that three of them are 98.39167 or lower. Because 3 of the 20 results (or 15%) are at least as extreme as the sample mean of 98.39167, we see that a sample mean such as 98.39167 is common for these circumstances. This suggests that a sample mean of 98.39167 is not *significantly* different from the assumed mean of 98.6. We would feel more confident in this conclusion if we had more sample results, so we could continue to randomly generate simulated samples until we feel quite confident in our thinking that a sample mean such as 98.39167 is not an unusual result. It can easily occur as the result of chance random variation. We therefore fail to reject the null hypothesis that the mean equals 98.6. There is not sufficient evidence to support the claim that the mean is less than 98.6. This is the same conclusion reached in Section 8–3 of this manual/workbook, where Minitab was used for a formal hypothesis test, but this simulation method uses a very different approach than the formal methods described in the textbook.

CHAPTER 8 EXPERIMENTS: Hypothesis Testing

Experiments 8–1 through 8–4 involve claims about proportions..

8–1. ***Photo–Cop Legislation*** Is there sufficient sample evidence to support a claim that the proportion of all adult Minnesotans opposed to photo–cop legislation is greater than 0.5? Use a 0.10 significance level to test the claim that the proportion is greater than 0.5. Sample data consists of $n = 829$ randomly selected adult Minnesotans with 51% opposed to photo–cop legislation.

Test statistic: _____   *P*–value: _____

Conclusion in your own words: _____

_____

8–2. ***Cloning Survey*** In a Gallup poll of 1012 randomly selected adults, 9% said that cloning of humans should be allowed. Use a 0.05 significance level to test the claim that less than 10% of all adults say that cloning of humans should be allowed. Can a newspaper run a headline that "less than 10% of adults believe that cloning of humans should be allowed"?

Test statistic: _____   *P*–value: _____

Conclusion in your own words: _____

_____

8–3. ***Store Checkout-Scanner Accuracy*** In a study of store checkout-scanners, 1234 items were checked and 20 of them were found to be overcharges (based on data from "UPC Scanner Pricing Systems: Are They Accurate?" by Goodstein, *Journal of Marketing,* Vol. 58). Use a 0.05 significance level to test the claim that with scanners, 1% of sales are overcharges. (Before scanners were used, the overcharge rate was estimated to be about 1%.) Based on these results, do scanners appear to help consumers avoid overcharges?

Test statistic: _____   *P*–value: _____

Conclusion in your own words: _____

_____

8–4.   ***Drug Testing of Job Applicants*** In 1990, 5.8% of job applicants who were tested for drugs failed the test. At the 0.01 significance level, test the claim that the failure rate is now lower if a simple random sample of 1520 current job applicants results in 58 failures (based on data from the American Management Association). Does the result suggest that fewer job applicants now use drugs?

Test statistic: _____   *P*–value: _____

Conclusion in your own words: _____

_____

Experiments 8–5 through 8–8 use original lists of sample data for testing claims about a population mean.

8–5.   ***Monitoring Lead in Air*** Listed below are measured amounts of lead (in micrograms per cubic meter or $\mu g/m^3$) in the air. The Environmental Protection Agency has established an air quality standard for lead: 1.5 $\mu g/m^3$. The measurements shown below were recorded at Building 5 of the World Trade Center site on different days immediately following the destruction caused by the terrorist attacks of September 11, 2001. After the collapse of the two World Trade buildings, there was considerable concern about the quality of the air. Use a 0.05 significance level to test the claim that the sample is from a population with a mean greater than the EPA standard of 1.5 $\mu g/m^3$. Is there anything about this data set suggesting that the assumption of a normally distributed population might not be valid?

5.40   1.10   0.42   0.73   0.48   1.10

Test statistic: _____   *P*–value: _____

Conclusion in your own words: _____

_____

8–6.   ***Treating Chronic Fatigue Syndrome*** Patients with chronic fatigue syndrome were tested, then retested after being treated with fludrocortisone. Listed below are the changes in fatigue after the treatment (based on data from "The Relationship Between Neurally Mediated Hypotension and the Chronic Fatigue Syndrom" by Bou–Holaigah, Rowe, Kan, and Calkins, *Journal of the American Medical Association,* Vol. 274, No. 12). A standard scale from −7 to +7 was used, with positive values representing improvements. Use a 0.01 significance level to test the claim that the mean change is positive. Does the treatment appear to be effective?

(*continued*)

6 5 0 5 6 7 3 3 2 6 5 5 0 6 3 4 3 7 0 4 4

Test statistic: _____   *P*–value: _____

Conclusion in your own words: _____

_____

8–7.   ***Olympic Winners*** Listed below are the winning times (in seconds) of men in the 100-meter dash for consecutive summer Olympic games, listed in order by row. Assuming that these results are sample data randomly selected from the population of all past and future Olympic games, test the claim that the mean time is less than 11 sec. What do you observe about the precision of the numbers? What extremely important characteristic of the data set is not considered in this hypothesis test? Do the results from the hypothesis test suggest that future winning times should be around 10.5 sec, and is such a conclusion valid?

12.0   11.0   11.0   11.2   10.8   10.8   10.8   10.6   10.8   10.3   10.3   10.3
10.4   10.5   10.2   10.0   9.95   10.14   10.06   10.25   9.99   9.92   9.96

Test statistic: _____   *P*–value: _____

Conclusion in your own words: _____

_____

_____

_____

8–8.   ***Nicotine in Cigarettes*** The Carolina Tobacco Company advertised that its best-selling nonfiltered cigarettes contain at most 40 mg of nicotine, but *Consumer Advocate* magazine ran tests of 10 randomly selected cigarettes and found the amounts (in mg) shown in the accompanying list. It's a serious matter to charge that the company advertising is wrong, so the magazine editor chooses a significance level of $\alpha = 0.01$ in testing her belief that the mean nicotine content is greater than 40 mg. Using a 0.01 significance level, test the editor's belief that the mean is greater than 40 mg.

47.3   39.3   40.3   38.3   46.3   43.3   42.3   49.3   40.3   46.3

Test statistic: _____   *P*–value: _____

Conclusion in your own words: _____

_____

Experiments 8–9 through 8–12 use summary statistics to test a claim about a population mean.

8–9.   ***Testing Wristwatch Accuracy*** Students of the author randomly selected 40 people and measured the accuracy of their wristwatches, with positive errors representing watches that are ahead of the correct time and negative errors representing watches that are behind the correct time. The 40 values have a mean of 117.3 sec and a standard deviation of 185.0 sec. Use a 0.01 significance level to test the claim that the population of all watches has a mean equal to 0 sec. What can be concluded about the accuracy of people's wristwatches?

Test statistic: _____   *P*–value: _____

Conclusion in your own words: _____

_____

8–10.   ***Conductor Life Span*** A *New York Times* article noted that the mean life span for 35 male symphony conductors was 73.4 years, in contrast to the mean of 69.5 years for males in the general population. Assuming that the 35 males have life spans with a standard deviation of 8.7 years, use a 0.05 significance level to test the claim that male symphony conductors have a mean life span that is greater than 69.5 years. Does it appear that male symphony conductors live longer than males from the general population? Why doesn't the experience of being a male symphony conductor cause men to live longer? (*Hint*: Are male symphony conductors born, or do they become conductors at a much later age?)

Test statistic: _____   *P*–value: _____

Conclusion in your own words: _____

_____

8–11.   ***Birth Weights*** In a study of the effects of prenatal cocaine use on infants, the following sample data were obtained for weights at birth: $n = 190$, $\bar{x} = 2700$ g, $s = 645$ g (based on data from "Cognitive Outcomes of Preschool Children With Prenatal Cocaine Exposure" by Singer, et al, *Journal of the American Medical Association*, Vol. 291, No. 20). Use a 0.01 significance level to test the claim that weights of babies born to cocaine users have a mean that is less than the mean of 3103 g for babies born to mothers who do not use cocaine. Based on the results, does it appear that birth weights are affected by cocaine use?

Test statistic: _____   *P*–value: _____

Conclusion in your own words: _____

_____

8–12. ***Baseballs*** In previous tests, baseballs were dropped 24 ft onto a concrete surface, and they bounced an average of 92.84 in. In a test of a sample of 40 new balls, the bounce heights had a mean of 92.67 in. and a standard deviation of 1.79 in. (based on data from Brookhaven National Laboratory and *USA Today*). Use a 0.05 significance level to determine whether there is sufficient evidence to support the claim that the new balls have bounce heights with a mean different from 92.84 in. Does it appear that the new baseballs are different?

Test statistic: _____   *P*–value: _____

Conclusion in your own words: _____

Experiments 8–13 through 8–16 involve claims about a standard deviation or variance.

8–13. ***Supermodel Weights*** Use a 0.01 significance level to test the claim that weights of female supermodels vary less than the weights of women in general. The standard deviation of weights of the population of women is 29 lb. Listed below are the weights (in pounds) of nine randomly selected supermodels.

125 (Taylor)   119 (Auermann)   128 (Schiffer) 128 (MacPherson)
119 (Turlington)   127 (Hall)   105 (Moss)   123 (Mazza)
115 (Hume)

Test statistic: _____   *P*–value: _____

Conclusion in your own words: _____

_____

8–14. ***Supermodel Heights*** Use a 0.05 significance level to test the claim that heights of female supermodels vary less than the heights of women in general. The standard deviation of heights of the population of women is 2.5 in. Listed below are the heights (in inches) of randomly selected supermodels (Taylor, Harlow, Mulder, Goff, Evangelista, Avermann, Schiffer, MacPherson, Turlington, Hall, Crawford, Campbell, Herzigova, Seymour, Banks, Moss, Mazza, Hume).

71   71   70   69   69.5   70.5   71   72   70
70   69   69.5   69   70   70   66.5   70   71

Test statistic: _____   *P*–value: _____

Conclusion in your own words: _____

_____

8–15. ***Volumes of Pepsi*** A new production manager claims that the volumes of cans of regular Pepsi have a standard deviation less than 0.10 oz. Use a 0.05 significance level to test that claim with the sample results listed in the "Weights and Volumes of Cola" data set in Appendix B of the textbook. What problems are caused by a mean that is not 12 oz? What problems are caused by a standard deviation that is too high?

Test statistic: _____   *P*–value: _____

Conclusion in your own words: _____

_____

8–16. ***Systolic Blood Pressure for Women*** Systolic blood pressure results from contraction of the heart. Based on past results from the National Health Survey, it is claimed that women have systolic blood pressures with a mean and standard deviation of 130.7 and 23.4, respectively. Use the systolic blood pressures of women listed in Data Set 1 in Appendix B and test the claim that the sample comes from a population with a standard deviation of 23.4.

Test statistic: _____   *P*–value: _____

Conclusion in your own words: _____

_____

The following experiments involve the simulation approach to hypothesis testing.

8–17. ***Hypothesis Testing with Simulations*** Suppose that we want to use a 0.05 significance level to test the claim that $p > 0.5$ and we have sample data summarized by $n = 100$ and $x = 70$. Use the following Minitab procedure for using a simulation method for testing the claim that $p > 0.5$.

We assume in the null hypothesis that $p = 0.5$. Working under the assumption that $p = 0.5$, we will generate different samples of size $n = 100$ until we have a sense for the likelihood of getting 70 or more successes. Begin by generating a sample of size 100 as follows: Select **Calc/Random Data/Integer**. In the dialog box, enter 100 for the number of rows, enter C1–C10 for the columns, then enter a minimum of 0 and a maximum of 1. After clicking **OK**, the result will be 10 columns, where each column represents a sample of 100 subjects with a probability of 0.5 of getting a 1 in each cell. Scroll through the data field to see that it simulates 10 samples of 100 subjects. Letting 1 represent a success, use **Stat/Display Descriptive Statistics** and find the number of samples having a sample proportion greater than 0.5. Enter that result here: _____

continued

What percentage of the simulated samples have a sample proportion greater than 0.5? Enter that result here: _____

What does the preceding result suggest about the likelihood of this event: "When the population proportion p is really equal to 0.5, a sample of size 100 is randomly selected and the sample proportion is greater than 0.5"

_____

Can we conclude that the probability of a sample proportion greater than 0.5 is less than or equal to the significance level of 0.05? _____

What is the final conclusion? _____

_____

The preceding results are based on 10 simulated samples. How can we increase the number of samples so that we have more confidence in our conclusions?

_____

_____

_____

8–18. ***Hypothesis Testing with Simulations*** Experiment 8–17 used a simulation approach to test a claim. Use a simulation approach for conducting the hypothesis test described in Experiment 8–4. Describe the procedure, results, and conclusions.

_____

_____

_____

_____

8–19. ***Hypothesis Testing with Simulations*** Refer to Experiment 8–6 in this manual/work-book. We want to test the claim that $\mu > 0$, and the given sample data can be summarized with these statistics: $n = 21$, $\bar{x} = 4.0000$, and $s = 2.1679$. Instead of conducting a formal hypothesis test, we will now consider another way of determining whether the sample mean of 4.0000 is significantly greater than the claimed value of 0. We will use a significance level of 0.10 (instead of 0.01 as in Experiment 8–6). Using a significance level of $\alpha = 0.10$, we will use this criterion:

The sample mean of 4.0000 is significantly greater than 0 if there is a 10% chance (or less) that the following event occurs: A sample mean of 4.0000 or greater is obtained when selecting a random sample of n = 21 values from a normally distributed population with mean μ = 0 and standard deviation σ = 2.1679.

Use **Calc/Random Data/Normal** to randomly generate a sample of 21 values from a normally distributed population with the assumed mean of 0 and a standard deviation of 2.1679. Find the mean of the generated sample and record it in the space below. Then generate another sample, and another sample, and continue until you have enough sample means to determine how often a result such as $\bar{x}$ = 4.0000 (or any other higher sample mean) will occur.

_____

_____

_____

Enter the number of generated samples:_____

How often was the sample mean equal to or greater than 4.0000? _____

What is the proportion of trials in which the sample mean was 4.0000 or higher? _____

Based on these results, what do you conclude about the claim that $\mu > 0$? Explain.

_____

_____

_____

8–20.   ***Hypothesis Testing with Simulations***   Repeat Experiment 8–7 in this manual/workbook by using a simulation method. (*Hint*: See Section 8–5 of this manual/workbook and see Experiment 8–19.) Describe the procedure used and describe your results and conclusion.

_____

_____

8–21. ***Hypothesis Testing with Simulations*** Assume that we have a random sample of 75 IQ scores from statistics professors, and the mean and standard deviation for this sample are 130 and 10, respectively. We want to use a 0.05 significance level to test the claim that the population of all statistics professors has a mean greater than 100, which is the mean IQ score for the general population.

Use **Calc/Random Data/Normal** to generate a sample of 75 IQ scores from a population having a mean of 100 and a standard deviation of 10. Enter the mean of the sample here: _____

Repeat the preceding step 19 more times and enter all 20 sample means here:

_____

_____

What is the proportion of sample means that are 130 or greater? _____

What does the preceding result suggest about the claim that the population of IQ scores of statistics professors is greater than 100? Explain.

_____

_____

_____

_____

9

Inferences from Two Samples

9-1 Working with Summary Statistics

As in Chapters 7 and 8 of this manual/workbook, Minitab sometimes requires lists of the original sample data. The table shown below shows the data requirements for Minitab Releases 13 and 14. If we know only the summary statistics, but Minitab requires lists of original data values, generate those lists as described in Section 6-5 of this manual/workbook.

	<u>Minitab Release 13</u>	<u>Minitab Release 14</u>
Two proportions	Use summary statistics.	Use summary statistics.
Two means (independent)	Use lists of sample data.	Use lists of sample data or summary statistics.
Two means (matched pairs)	Use lists of sample data.	Use lists of sample data or summary statistics of the differences.
Two variances	Use lists of sample data.	Use lists of sample data or summary statistics.

9-2 Two Proportions

The textbook makes the point that the section discussing inferences involving two proportions is one of the most important sections in the book because the main objective is to provide methods for dealing with two sample proportions — a situation that is very common in real applications."

Using Minitab, we will first identify the number of successes x_1 and the sample size n_1 for the first sample, and identify x_2 and n_2 for the second sample. Sample data often consist of sample proportions instead of the actual numbers of successes, so be sure to read the textbook carefully for a way to determine the number of successes. From $\hat{p}_1 = x_1/n_1$, we know that $x_1 = n_1 \cdot \hat{p}_1$ so that x_1 can be found by multiplying the sample size for the first sample by the sample proportion expressed in decimal form. After identifying n_1, x_1, n_2, and x_2, proceed with Minitab as follows.

Minitab Procedure for Inferences (Hypothesis Tests and Confidence Intervals) for Two Proportions

To conduct hypothesis tests or construct confidence intervals for two population proportions, use the following procedure.

　　1.　　For each sample, find the sample size n and the number of successes x.

2.   Select **Stat**, then **Basic Statistics**, then **2 Proportions**.

3.   You will now get a dialog box.

Select the option of "Summarized data" and enter the number of trials and the number of successes (or "events") for each of the two samples.

4.   Click on the **Options** bar and proceed to enter the confidence level (enter 95 for a 0.05 significance level), enter the claimed value of the difference (usually 0), and enter the format for the alternative hypothesis.

5.   While still in the Options dialog box, refer to the box with the label of "Use pooled estimate of p for test."

- If testing a hypothesis, click on that box (because you want to use a pooled estimate of p).

- If constructing a confidence interval, do *not* click on that box (because you do not want to use a pooled estimate of p).

As an illustration, consider this example:

Some patients suffering from carpal tunnel syndrome were treated with surgery, while others were treated with splints. Success of the treatment was determined one year later, and the results are given in the table below. In a journal article about the trial, authors claimed that "Treatment with open carpal tunnel release surgery resulted in better outcomes than treatment with wrist splinting for patients with CTS (carpal tunnel syndrome)." Do the sample results really support the claim that surgery is better?

Treatments of Carpal Tunnel Syndrome

	Surgery	Splint
Success after one year	67	60
Total number treated	73	83
Success Rate	**92%**	**72%**

For notation purposes, we stipulate that sample 1 is the group treated with surgery, and sample 2 is the group treated with splints. We can now proceed to use the above Minitab procedure. After selecting **Stat**, **Basic Statistics**, then **2 Proportions**, we make the required entries in the dialog boxes as shown below. Because the form of the claim is that the success rate for the surgery treatment group is *greater than* the success rate for those treated with splints, click on **Options** and make the entries as shown in the Options box below.

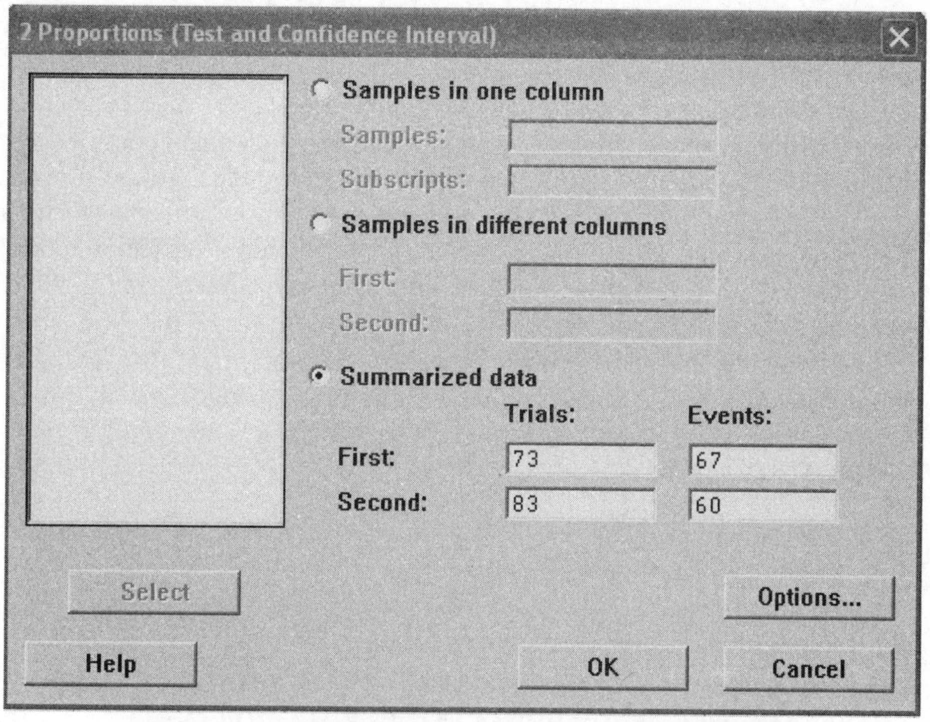

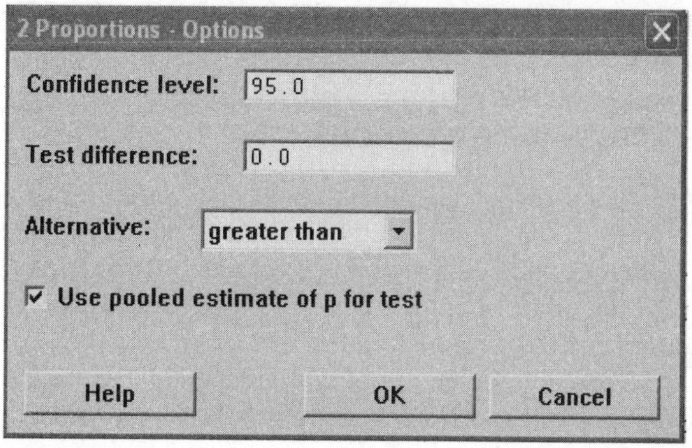

The Minitab results include the test statistic and P–value. The display shows a P–value of 0.001, indicating that we should reject the null hypothesis and support the claim that the surgery treatment group has a greater success rate than the splint treatment group.

```
Sample   X    N   Sample p
1        67   73  0.917808
2        60   83  0.722892
```

```
Difference = p (1) - p (2)
Estimate for difference:  0.194917
95% lower bound for difference:  0.0983474
Test for difference = 0 (vs > 0):  Z = 3.12   P-Value = 0.001
```

Confidence Intervals: The above Minitab display includes a one–sided confidence interval, but it is based on the pooling of the two sample proportions. (See Step 5 in the preceding Minitab procedure. Note that we check the box for pooling the sample proportions only if we are testing a hypothesis. If we are constructing a confidence interval, we should *not* check the box for pooling the sample proportions.) If we use the given sample data for constructing a 90% confidence interval estimate of $p_1 - p_2$, we get $0.0983 < p_1 - p_2 < 0.291$ from the results shown below. Remember, when clicking on the Options box, select "not equal to" for the alternative hypothesis, and do *not* click on the box for pooling the sample standard deviations.

```
Sample   X    N   Sample p
1        67   73  0.917808
2        60   83  0.722892
```

```
Difference = p (1) - p (2)
Estimate for difference:  0.194917
90% CI for difference:  (0.0983474, 0.291486)
Test for difference = 0 (vs not = 0):  Z = 3.32   P-Value = 0.001
```

9-3 Two Means: Independent Samples

The textbook notes that two samples are **independent** if the sample values selected from one population are not related to or somehow paired or matched with the sample values selected from the other population. If there is some relationship so that each value in one sample is paired with a corresponding value in the other sample, the samples are **dependent.** Dependent samples are often referred to as **matched pairs,** or **paired samples.**

When testing a claim about the means of two independent samples, or when constructing a confidence interval estimate of the difference between the means of two independent samples, the textbook describes procedures based on the requirement that the two population standard deviations σ_1 and σ_2 are not known, and there is no assumption that $\sigma_1 = \sigma_2$. We focus on the first of the following three cases, but the other two cases are discussed briefly:

1. σ_1 and σ_2 are not known and are not assumed to be equal.
2. σ_1 and σ_2 are known.
3. It is assumed that $\sigma_1 = \sigma_2$.

The first of these three cases is the main focus of the textbook, and Minitab handles that first case but, instead of using "the smaller of $n_1 - 1$ and $n_2 - 1$" for the number of degrees of freedom, Minitab calculates the better number of degrees of freedom given as Formula 9-1 in the *Elementary Statistics,* 10th edition.

Minitab also provides the option of assuming that $\sigma_1 = \sigma_2$, but there is no option for assuming that σ_1 and σ_2 are *known*. That is, Minitab allows us to deal with cases 1 and 3 listed above, but not case 2. We describe the Minitab procedure, then consider an example.

Minitab Procedure for Inferences (Hypothesis Tests and Confidence Intervals) about Two Means: Independent Samples

Use the following procedure for both hypothesis tests and confidence intervals.

1. Identify the values of the summary statistics, or enter the original sample values in columns C1 and C2. (If using Minitab Release 13 or earlier and the original sample values are not know, generate two samples with the same sample sizes, means, and standard deviations. See Section 6-5. Store the results in columns C1 and C2.)

2. Click on the main menu item of **Stat**, then select **Basic Statistics**, then **2-Sample t**.

3. Proceed to make the entries in the dialog box.

    -   Do *not* check the box labeled "Assume equal variances" (unless there is a sound justification for assuming that $\sigma_1 = \sigma_2$). Be sure that the box remains unchecked. We are not assuming equal variances.

    -   Click on the **Options** box and enter the confidence level (use 95 for a 0.05 significance level), enter the claimed difference (usually 0), and select the format of the alternative hypothesis. Click **OK** for the Options box, then click **OK** for the 2-Sample t box.

As an illustration, consider these sample data and the claim given on the following page:

Discrimination Based on Age The Revenue Commissioners in Ireland conducted a contest for promotion. The ages of the unsuccessful and successful applicants are given below (based on data from "Debating the Use of Statistical Evidence in Allegations of Age Discrimination" by Barry and Boland, *The American Statistician,* Vol. 58, No. 2). Some of the applicants who were unsuccessful in getting the promotion charged that the competition involved discrimination based on age. Treat the data as samples from larger populations and use a 0.05 significance level to test the claim that the unsuccessful applicants are from a population with a greater mean age than the mean age of successful applicants.

Ages of Unsuccessful Applicants	Ages of Successful Applicants
34 37 37 38 41 42 43 44 44 45	27 33 36 37 38 38 39 42 42 43
45 45 46 48 49 53 53 54 54 55	43 44 44 44 45 45 45 45 46 46
56 57 60	47 47 48 48 49 49 51 51 52 54

To test the given claim using Minitab, first enter the data from the first sample in column C1, then enter the data for the second sample in column C2. Select **Stat**, **Basic Statistics**, then **2–Sample t**. After following the above procedure for testing the claim, the Minitab results will be as shown below. The test statistic is $t = 1.63$ and the *P*–value is 0.055, so we fail to reject the null hypothesis. There is not sufficient evidence to support the claim that the unsuccessful applicants have a greater mean age than successful applicants.

```
Two-sample T for C1 vs C2

      N    Mean   StDev  SE Mean
C1   23   46.96    7.22      1.5
C2   30   43.93    5.88      1.1


Difference = mu (C1) - mu (C2)
Estimate for difference:  3.02319
95% lower bound for difference:  -0.08926
T-Test of difference = 0 (vs >): T-Value = 1.63   P-Value = 0.055
DF = 41
```

9-4 Matched Pairs

The textbook describes methods for testing hypotheses and constructing confidence interval estimates of the differences between samples consisting of *matched pairs*. Here is the Minitab procedure:

**Minitab Procedure for Inferences (Hypothesis Tests and Confidence Intervals)
about Two Means: Matched Pairs**

To conduct hypothesis tests or construct confidence intervals based on data consisting of matched pairs, use the following procedure.

1. Enter the values of the first sample in column C1.
    Enter the values of the second sample in column C2.

2. Select **Stat, Basic Statistics, Paired t**.

3. Make these entries and selections in the dialog box that pops up:
    - Enter C1 in the box labeled "First sample.
    - Enter C2 in the box labeled "Second sample."
    - Click the **Options** bar and proceed to enter the confidence level (enter 95 for a 0.05 significance level), enter the claimed value of the difference μ_d, and select the form of the alternative hypothesis. Click **OK**, then click **OK** for the main dialog box.

See the following example.

Are Forecast Temperatures Accurate? The table below consists of five actual low temperatures and the corresponding low temperatures that were predicted five days earlier. The data consist of matched pairs, because each pair of values represents the same day. The forecast temperatures appear to be very different from the actual temperatures, but is there sufficient evidence to conclude that the mean difference is not zero? Use a 0.05 significance level to test the claim that there is a difference between the actual low temperatures and the low temperatures that were forecast five days earlier.

Actual and Forecast Temperature

Actual Low	54	54	55	60	64
Low Forecast Five Days Earlier	56	57	59	56	64
Difference $d =$ Actual − Predicted	−2	−3	−4	4	0

Here is the Minitab display that results from the preceding hypothesis test. It shows a *P*-value of 0.519, so we fail to reject the null hypothesis. There isn't sufficient evidence to support the claim that the mean difference is not zero.

```
95% CI for mean difference: (-4.92649, 2.92649)
T-Test of mean difference = 0 (vs not = 0):
T-Value = -0.71   P-Value = 0.519
```

9-5 Two Variances

The textbook describes the use of the F distribution in testing a claim that two populations have the same variance (or standard deviation). Minitab is capable of conducting such an F test.

Minitab Procedure for Inferences (Hypothesis Tests and Confidence Intervals) for Two Standard Deviations or Two Variances

1. Either obtain the summary statistics for each sample, or enter the individual sample values in two columns. (If using Minitab Release 13 or earlier, you must have the original lists of raw data. If you know only the summary statistics, generate two lists of sample data as described in Section 6-5.)

2. Select **Stat**, then **Basic Statistics**, then **2 Variances**.

3. A dialog box will appear. Either select the option of "Samples in different columns" and proceed to enter the column names, or select "Summarized data" and proceed to enter the summary statistics (if using Minitab Release 14).

4. Click on the **Options** bar and enter the confidence level. (Enter 95 for a 0.05 significance level.) Click **OK**, then click **OK** in the main dialog box.

The textbook gives an example using a 0.05 significance level in testing the claim that the two populations of weights of regular Coke and regular Pepsi have the same standard deviation. (The weights are in the data set "Weights and Volumes of Cola" from Appendix B in the textbook.) The following Minitab results of the F-test include a P-value of 0.108, which is greater than the significance level, so we fail to reject the null hypothesis of equal variances. It appears that the two standard deviations do not differ by a significant amount.

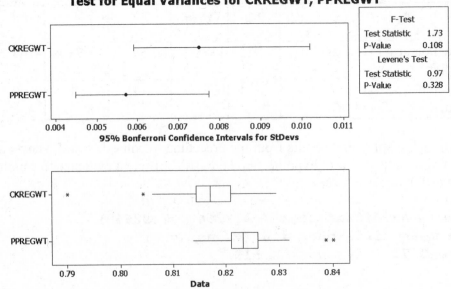

CHAPTER 9 EXPERIMENTS: Inferences from Two Samples

9–1.   *E–Mail and Privacy* A survey of 436 workers showed that 192 of them said that it was seriously unethical to monitor employee e–mail. When 121 senior–level bosses were surveyed, 40 said that it was seriously unethical to monitor employee e–mail (based on data from a Gallup poll). Use a 0.05 significance level to test the claim that for those saying that monitoring e–mail is seriously unethical, the proportion of employees is greater than the proportion of bosses.

Test statistic: _____   *P*–value: _____

Conclusion in your own words: _____

_____

9–2.   *E–Mail and Privacy* Refer to the sample data given in Experiment 9–1 and construct a 90% confidence interval estimate of the difference between the two population proportions. Is there a substantial gap between the employees and bosses?

_____

_____

9–3.   *Exercise and Coronary Heart Disease* In a study of women and coronary heart disease, the following sample results were obtained: Among 10,239 women with a low level of physical activity (less than 200 kcal/wk), there were 101 cases of coronary heart disease. Among 9877 women with physical activity measured between 200 and 600 kcal/wk, there were 56 cases of coronary heart disease (based on data from "Physical Activity and Coronary Heart Disease in Women" by Lee, Rexrode, et al, *Journal of the American Medical Association,* Vol. 285, No. 11). Construct a 90% confidence interval estimate for the difference between the two proportions. Does the difference appear to be substantial? Does it appear that physical activity corresponds to a lower rate of coronary heart disease?

_____

_____

9–4.   *Exercise and Coronary Heart Disease* Refer to the sample data in Experiment 9–3 and use a 0.05 significance level to test the claim that the rate of coronary heart disease is higher for women with the lower levels of physical activity.

Test statistic: _____   *P*–value: _____

(*continued*)

Conclusion in your own words: _____

_____

9-5. ***Instant Replay in Football*** In the 2000 football season, 247 plays were reviewed by officials using instant video replays, and 83 of them resulted in reversal of the original call. In the 2001 football season, 258 plays were reviewed and 89 of them were reversed (based on data from "Referees Turn to Video Aid More Often" by Richard Sandomir, *New York Times*). Is there a significant difference in the two reversal rates?

Test statistic: _____   *P*-value: _____

Conclusion in your own words: _____

_____

Does it appear that the reversal rate was the same in both years? _____

9-6. ***Effectiveness of Smoking Bans*** The Joint Commission on Accreditation of Healthcare Organizations mandated that hospitals ban smoking by 1994. In a study of the effects of this ban, subjects who smoke were randomly selected from two different populations. Among 843 smoking employees of hospitals with the smoking ban, 56 quit smoking one year after the ban. Among 703 smoking employees from workplaces without a smoking ban, 27 quit smoking a year after the ban (based on data from "Hospital Smoking Bans and Employee Smoking Behavior" by Longo, Brownson, et al, *Journal of the American Medical Association*, Vol. 275, No. 16). Is there a significant difference between the two proportions at a 0.05 significance level?

Test statistic: _____   *P*-value: _____

Conclusion in your own words: _____

_____

Is there a significant difference between the two proportions at a 0.01 significance level?

Test statistic: _____   *P*-value: _____

Conclusion in your own words: _____

_____

Does it appear that the ban had an effect on the smoking quit rate? _____

9–7. ***Testing Effectiveness of Vaccine*** In a *USA Today* article about an experimental nasal spray vaccine for children, the following statement was presented: "In a trial involving 1602 children only 14 (1%) of the 1070 who received the vaccine developed the flu, compared with 95 (18%) of the 532 who got a placebo." The article also referred to a study claiming that the experimental nasal spray "cuts children's chances of getting the flu." Is there sufficient sample evidence to support the stated claim?

Test statistic: _____ *P*–value: _____

Conclusion in your own words: _____

_____

In Experiments 9–8 through 9–15, assume that the two samples are independent simple random samples selected from normally distributed populations. Do not assume that the population standard deviations are equal.

9–8. ***Axial Loads of Cans*** Refer to the CANS data set in Appendix B in the textbook. (The Minitab worksheet name is CANS.) Is there a significant difference between the axial loads of cans with thickness 0.0109 in. and cans with 0.0111 in. thickness?

Test statistic: _____ *P*–value: _____

Conclusion in your own words: _____

_____

9–9. ***Weights of M&Ms*** Refer to the weights of red M&Ms and green M&Ms listed in Appendix B in the textbook. (The Minitab worksheet name is M&M.) Is there a significant difference between the weights of red M&Ms and the weights of green M&Ms?

Test statistic: _____ *P*–value: _____

Conclusion in your own words: _____

_____

9–10. ***BMI of Men and Women*** Refer to Data Set 1 in Appendix B of the textbook and test the claim that the mean body mass index (BMI) of men is equal to the mean body mass index of women.

Test statistic: _____ *P*–value: _____

Conclusion in your own words: _____

_____

9-11. ***Pulse Rates of Men and Women***   Refer to Data Set 1 in Appendix B of the textbook and test the claim that the mean pulse rate of men is equal to the mean pulse rate of women.

Test statistic: _____ P-value: _____

Conclusion in your own words: _____

_____

9-12. ***Hypothesis Test for Effect of Marijuana Use on College*** Students Many studies have been conducted to test the effects of marijuana use on mental abilities. In one such study, groups of light and heavy users of marijuana in college were tested for memory recall, with the results given below (based on data from "The Residual Cognitive Effects of Heavy Marijuana Use in College Students" by Pope and Yurgelun–Todd, Journal of the American Medical Association, Vol. 275, No. 7). Use a 0.01 significance level to test the claim that the population of heavy marijuana users has a lower mean than the light users.

Items sorted correctly by light marijuana users: n = 64, $\bar{x}$ = 53.3, s = 3.6
Items sorted correctly by heavy marijuana users: n = 65, $\bar{x}$ = 51.3, s = 4.5

Test statistic: _____ P-value: _____

Conclusion in your own words: _____

_____

Based on these results, should marijuana use be of concern to college students?

_____

9-13. ***Confidence Interval for Effects of Marijuana Use on College Students*** Refer to the sample data used in Experiment 9–12 and construct a 98% confidence interval for the difference between the two population means. Does the confidence interval include zero? What does the confidence interval suggest about the equality of the two population means?

_____

9-14. ***Hypothesis Test for Magnet Treatment of Pain***   People spend huge sums of money (currently around $5 billion annually) for the purchase of magnets used to treat a wide variety of pains. Researchers conducted a study to determine whether magnets are effective in treating back pain. Pain was measured using the visual analog scale, and the results given below are among the results obtained in the study (based on data from "Bipolar Permanent Magnets for the Treatment of Chronic Lower Back Pain: A Pilot Study" by Collacott, Zimmerman, White, and Rindone, Journal of the American Medical Association, Vol. 283, No. 10). Use a 0.05 significance level to test the claim that those treated with magnets have a greater reduction in pain than those given a sham treatment (similar to a placebo).

Reduction in pain level after magnet treatment: n = 20, $\bar{x}$ = 0.49, s = 0.96

Reduction in pain level after sham treatment: n = 20, $\bar{x}$ = 0.44, s = 1.4

Test statistic: _____   *P*–value: _____

Conclusion in your own words: _____

_____

Does it appear that magnets are effective in treating back pain? Is it valid to argue that magnets might appear to be effective if the sample sizes are larger?

_____

_____

9–15. ***Confidence Interval for Magnet Treatment of Pain***   Refer to the sample data from Experiment 9–14 and construct a 90% confidence interval estimate of the difference between the mean reduction in pain for those treated with magnets and the mean reduction in pain for those given a sham treatment. Based on the result, does it appear that the magnets are effective in reducing pain?

_____

9–16. ***Self–Reported and Measured Female Heights***   As part of the National Health and Nutrition Examination Survey conducted by the Department of Health and Human Services, self–reported heights and measured heights were obtained for females aged 12–16. Listed below are sample results.

Reported height	53	64	61	66	64	65	68	63	64	64	64	67
Measured height	58.1	62.7	61.1	64.8	63.2	66.4	67.6	63.5	66.8	63.9	62.1	68.5

Is there sufficient evidence to support the claim that there is a difference between self–reported heights and measured heights of females aged 12–16? Use a 0.05 significance level.

Test statistic: _____   *P*–value: _____

(continued)

Conclusion in your own words: _____

_____

Construct a 95% confidence interval estimate of the mean difference between reported heights and measured heights. Interpret the resulting confidence interval, and comment on the implications of whether the confidence interval limits contain 0.

_____

_____

9–17. ***Self–Reported and Measured Male Heights*** As part of the National Health and Nutrition Examination Survey conducted by the Department of Health and Human Services, self–reported heights and measured heights were obtained for males aged 12–16. Listed below are sample results.

Reported height	68	71	63	70	71	60	65	64	54	63	66	72
Measured height	67.9	69.9	64.9	68.3	70.3	60.6	64.5	67.0	55.6	74.2	65.0	70.8

Is there sufficient evidence to support the claim that there is a difference between self–reported heights and measured heights of males aged 12–16? Use a 0.05 significance level.

Test statistic: _____   *P*–value: _____

Conclusion in your own words: _____

_____

Construct a 95% confidence interval estimate of the mean difference between reported heights and measured heights. Interpret the resulting confidence interval, and comment on the implications of whether the confidence interval limits contain 0.

_____

_____

9–18. ***Effectiveness of SAT Course*** Refer to the data in the table that lists SAT scores before and after the sample of 10 students took a preparatory course (based on data from the College Board and "An Analysis of the Impact of Commercial Test Preparation Courses on SAT Scores," by Sesnowitz, Bernhardt, and Knain, *American Educational Research Journal,* Vol. 19, No. 3.)

(continued)

Student	A	B	C	D	E	F	G	H	I	J
SAT score before course (x)	700	840	830	860	840	690	830	1180	930	1070
SAT score after course (y)	720	840	820	900	870	700	800	1200	950	1080

Is there sufficient evidence to conclude that the preparatory course is effective in raising scores? Use a 0.05 significance level.

Test statistic: _____ P–value: _____

Conclusion in your own words: _____

_____

Construct a 95% confidence interval estimate of the mean difference between the before and after scores. Write a statement that interprets the resulting confidence interval.

_____

9–19. ***Before/After Treatment Results*** Captopril is a drug designed to lower systolic blood pressure. When subjects were tested with this drug, their systolic blood pressure readings (in mm of mercury) were measured before and after the drug was taken, with the results given in the accompanying table (based on data from "Essential Hypertension: Effect of an Oral Inhibitor of Angiotensin-Converting Enzyme," by MacGregor et al., *British Medical Journal,* Vol. 2).

Subject	A	B	C	D	E	F	G	H	I	J	K	L
Before	200	174	198	170	179	182	193	209	185	155	169	210
After	191	170	177	167	159	151	176	183	159	145	146	177

Use the sample data to construct a 99% confidence interval for the mean difference between the before and after readings.

_____

Is there sufficient evidence to support the claim that captopril is effective in lowering systolic blood pressure?

_____

9-20. ***Rainfall on Weekends***   *USA Today* and other newspapers reported on a study that supposedly showed that it rains more on weekends. The study referred to areas on the east coast near the ocean. A data set in Appendix B of the textbook lists the rainfall amounts in Boston for one year. (The Minitab worksheet name is BOSTRAIN.) The 52 rainfall amounts for Wednesday have a mean of 0.0517 in. and a standard deviation of 0.1357 in. The 52 rainfall amounts for Sunday have a mean of 0.0677 in. and a standard deviation of 0.2000 in.

Assuming that we want to use the methods in the textbook to test the claim that Wednesday and Sunday rainfall amounts have the same standard deviation, identify the F test statistic, P-value, and conclusion. Use a 0.05 significance level.

Test statistic: _____ P-value: _____

Conclusion in your own words: _____

_____

Consider the prerequisite of normally distributed populations. Instead of constructing histograms or normal quantile plots, simply examine the numbers of days with no rainfall. Are Wednesday rainfall amounts normally distributed? Are Sunday rainfall amounts normally distributed? What can be concluded from these results?

_____

_____

9-21. ***Tobacco and Alcohol Use in Animated Children's Movies*** A data set in Appendix B of the textbook lists times (in seconds) that animated children's movies show tobacco use and alcohol use. The 50 times of tobacco use have a mean of 57.4 sec and a standard deviation of 104.0 sec. The 50 times of alcohol use have a mean of 32.46 sec and a standard deviation of 66.3 sec. Assuming that we want to use the methods of this section to test the claim that the times of tobacco use and the times of alcohol use have different standard deviations, identify the F test statistic, P-value, and conclusion. Use a 0.05 significance level.

Test statistic: _____ P-value: _____

Conclusion in your own words: _____

_____

Consider the prerequisite of normally distributed populations. Instead of constructing histograms or normal quantile plots, simply examine the numbers of movies showing no tobacco or alcohol use. Are the times for tobacco use normally distributed? Are the times for alcohol use normally distributed? What can be concluded from these results?

_____

9–22. ***Calcium and Blood Pressure*** Sample data were collected in a study of calcium supplements and their effects on blood pressure. A placebo group and a calcium group began the study with blood pressure measurements (based on data from "Blood Pressure and Metabolic Effects of Calcium Supplementation in Normotensive White and Black Men," by Lyle et al., *Journal of the American Medical Association,* Vol. 257, No. 13). Sample values are listed below. At the 0.05 significance level, test the claim that the two sample groups come from populations with the same standard deviation.

Placebo:	124.6	104.8	96.5	116.3	106.1	128.8	107.2	123.1
	118.1	108.5	120.4	122.5	113.6			
Calcium:	129.1	123.4	102.7	118.1	114.7	120.9	104.4	116.3
	109.6	127.7	108.0	124.3	106.6	121.4	113.2	

Test statistic: _____   *P*–value: _____

Conclusion in your own words: _____

_____

If the experiment requires groups with equal standard deviations, are these two groups acceptable? _____

9–23. ***Homerun Distances*** Refer to the data set in Appendix B from the textbook that lists distances of home runs. (The Minitab worksheet name is HOMERUNS.) Use a 0.05 significance level to test the claim that distances of homeruns hit by Barry Bonds and Sammy Sosa have the same variation.

Test statistic: _____   *P*–value: _____

Conclusion in your own words: _____

_____

9–24. ***Homerun Distances*** Refer to the data set in Appendix B from the textbook that lists distances of home runs. (The Minitab worksheet name is HOMERUNS.) Use a 0.05 significance level to test the claim that distances of homeruns hit by Mark McGwire and Sammy Sosa have the same variation.

Test statistic: _____   *P*–value: _____

Conclusion in your own words: _____

_____

10

Correlation
and
Regression

10-1 Scatterplot

The Correlation and Regression chapter in the textbook introduces the basic concepts of linear correlation and regression. The basic objective is to use paired sample data to determine whether there is a relationship between two variables and, if so, identify what the relationship is. Consider the paired sample data in the table below. Each duration time (in seconds) is the duration of an eruption of the Old Faithful geyser, and the corresponding interval after is the time (in minutes) to the next eruption. We want to determine whether there is a relationship between duration and the time interval after an eruption. If such a relationship exists, we want to identify it with an equation so that we can predict the time of the next eruption.

Old Faithful Geyser

Duration	240	120	178	234	235	269	255	220
Interval After	92	65	72	94	83	94	101	87

The Minitab procedure for constructing a scatterplot was given in Section 2-6 of this manual/workbook. Enter the paired data in columns C1 and C2, then select **Graph**, then **Plot**.

10-2 Correlation

The textbook describes the linear correlation coefficient r as a measure of the strength of the linear relationship between two variables. Minitab can compute the value of r for paired data. Instead of doing the complicated manual calculations, follow these steps to let Minitab calculate r.

Minitab Procedure for Calculating the Linear Correlation Coefficient r

1.   Enter the x values in column C1, and enter the corresponding y values in column C2. It is also wise to enter the names of the variables in the empty cells located immediately above the first row of data values.

2.   Select **Stat**, then **Basic Statistics**, then **Correlation**.

3.   In the dialog box, enter the column labels of C1 C2, or enter the variable names, then click **OK**.

Shown below is the Minitab display that results from using the data in the above table. The variable names of Duration and IntAfter were also entered. Because the P-value of 0.001 is less than the significance level of 0.05, we conclude that there is a significant linear correlation between the duration and the time interval after an eruption.

Correlations: Duration, IntAfter

```
Pearson correlation of Duration and IntAfter = 0.926
P-Value = 0.001
```

10-3 Regression

The textbook discusses the important topic of regression. Given a collection of paired sample data, the **regression equation**

$$\hat{y} = b_0 + b_1x$$

describes the relationship between the two variables. The graph of the regression equation is called the **regression line** (or *line of best fit*, or *least-squares line*). The regression equation expresses a relationship between x (called the *independent variable* or *predictor variable*) and y (called the *dependent variable* or *response variable*). The y-intercept of the line is b_0 and the slope is b_1. Given a collection of paired data, we can find the regression equation as follows.

Minitab Procedure for Finding the Equation of the Regression Line

1.   Enter the x values in column C1, and enter the corresponding y values in column C2. Also enter the names of the variables in the empty cells located immediately above the first row of data values.

2.   Select **Stat**, then **Regression**, then **Regression** once again.

3.   Make these entries in the dialog box:

   - Enter C2 (the y values) in the box for the "Response" variable.

   - Enter C1 (the x values) in the box for the "Predictor" variable.

   - Click on the **Options** bar and click on the **"Fit intercept"** box.

   - Click on the **Results** bar and select the second option that includes the regression equation, s, and R-Squared.

   - Click **OK**.

Referring to the sample data in the preceding table, if you use the response variable of IntAfter [for the time interval (in minutes) after an eruption] and the predictor variable of duration time (in seconds), the Minitab display will be as follows.

Regression Analysis: IntAfter versus Duration

```
The regression equation is
IntAfter = 34.8 + 0.234 Duration


Predictor      Coef   SE Coef      T       P
Constant     34.770     8.732   3.98   0.007
Duration    0.23406   0.03908   5.99   0.001


S = 4.97392    R-Sq = 85.7%    R-Sq(adj) = 83.3%

Analysis of Variance

Source            DF        SS       MS       F      P
Regression         1    887.56   887.56   35.88  0.001
Residual Error     6    148.44    24.74
Total              7   1036.00
```

This display includes the regression equation, which we express as $\hat{y} = 34.8 + 0.234x$. The Minitab display also includes "R-sq =85.7%," which is the coefficient of determination defined in the textbook; that value indicates that 85.7% of the variation in time intervals after an eruption can be explained by the duration of the eruption, and 14.3% of the variation in time intervals after eruptions remains unexplained. The display also includes other items discussed later in the textbook.

In addition to the equation of the regression line, we can also get a graph of the regression line superimposed on the scatter diagram.

Minitab Procedure for Graph of Regression Line

1. Enter the sample data in columns C1 and C2, and also enter the names of the variables in the empty cells immediately above the first row of data values.

2. Select **Stat**, **Regression**, then **Fitted line plot**.

3. Complete the dialog box by entering the response (y) variable, the predictor (x) variable, and select the **Linear** model. Click **OK**.

Shown below is the graph of the regression line for the data in the preceding table. We can see that the regression line fits the points quite well.

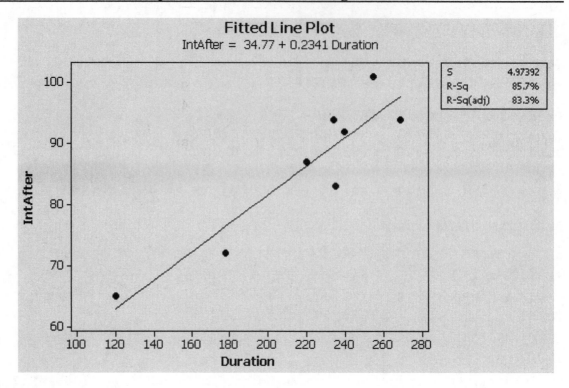

10-4 Predictions

We can use the regression equation for making predictions. For example, suppose that we want to predict the time interval (in minutes) after an eruption (y) given that the duration of an eruption is $x = 180$ sec. To find the predicted value of y given $x = 180$, follow the same procedure for finding the equation of the regression line as given in Section 10-3 of this manual/wokbook (using **Stat/Regression/Regression**) with this addition:

    In the **Options** box: 1.   Enter the x value(s) in the box labeled "Prediction intervals for new observations."

                                2.   Enter the desired confidence level, such as 95 (for 95% confidence).

If these two steps are followed by using $x = 180$, the following display will also be included in addition to the regression equation results.

```
Predicted Values for New Observations

New
Obs    Fit   SE Fit      95% CI           95% PI
 1    76.90    2.32   (71.21, 82.59)   (63.47, 90.33)

Values of Predictors for New Observations

New
Obs   Duration
 1       180
```

The above results show that using the regression equation of $\hat{y} = 34.8 + 0.234x$, the best predicted value of y (given $x = 180$ sec) is 76.90 min, and a 95% prediction interval is

$$63.47 \text{ min} < y < 90.33 \text{ min}$$

The values shown in the display for "95% C.I." are confidence interval limits for the mean time interval after an eruption for $x = 180$ sec.

10-5 Multiple Regression

The textbook discusses multiple regression, and Minitab can provide multiple regression results. Once a collection of sample data has been entered, you can easily experiment with different combinations of columns (variables) to find the combination that is best. Here is the Minitab procedure.

Minitab Procedure for Multiple Regression

1.   Enter the sample data in columns C1, C2, C3, . . . Also enter the names of the variables in the empty cells located immediately above the first row of sample values.

2.   Click on **Stat**, **Regression**, then **Regression** once again.

3.   Make these entries in the dialog box:

   • Enter the column containing the y values in the box for the "Response" variable.
   • In the box for the "Predictor" variables, enter the columns that you want included for the independent variables (the x variables).
   • Click on the **Options** bar and click on the "**Fit intercept**" box.
   • Click on the **Results** bar and select the second option that includes the regression equation, s, and R-Squared, then click **OK**.

As an example, the table below includes measurements obtained from anesthetized bears. If you enter the sample data included in the table and select column C1 as the response variable and columns C3 and C6 as predictor variables, the Minitab results will be as shown below the table. The results correspond to this multiple regression equation:

$$\hat{y} = -374 + 18.8x_3 + 5.87x_6$$

The results also include the adjusted coefficient of determination (75.9%) as well as a P-value (0.012) for overall significance of the multiple regression equation.

Data from Anesthetized Male Bears

Var.	Minitab Column	Name				Sample Data				
y	C1	WEIGHT	80	344	416	348	262	360	332	34
x_2	C2	AGE	19	55	81	115	56	51	68	8
x_3	C3	HEADLEN	11.0	16.5	15.5	17.0	15.0	13.5	16.0	9.0
x_4	C4	HEADWDTH	5.5	9.0	8.0	10.0	7.5	8.0	9.0	4.5
x_5	C5	NECK	16.0	28.0	31.0	31.5	26.5	27.0	29.0	13.0
x_6	C6	LENGTH	53.0	67.5	72.0	72.0	73.5	68.5	73.0	37.0
x_7	C7	CHEST	26	45	54	49	41	49	44	19

Here is the Minitab screen showing the entry of the values in the above table:

→	C1	C2	C3	C4	C5	C6	C7
	WEIGHT	AGE	HEADLEN	HEADWTH	NECK	LENGTH	CHEST
1	80	19	11.0	5.5	16.0	53.0	26
2	344	55	16.5	9.0	28.0	67.5	45
3	416	81	15.5	8.0	31.0	72.0	54
4	348	115	17.0	10.0	31.5	72.0	49
5	262	56	15.0	7.5	26.5	73.5	41
6	360	51	13.5	8.0	27.0	68.5	49
7	332	68	16.0	9.0	29.0	73.0	44
8	34	8	9.0	4.5	13.0	37.0	19

```
The regression equation is
WEIGHT = - 374 + 18.8 HEADLEN + 5.87 LENGTH
```

Predictor	Coef	SE Coef	T	P
Constant	-374.3	134.1	-2.79	0.038
HEADLEN	18.82	23.15	0.81	0.453
LENGTH	5.875	5.065	1.16	0.299

```
S = 68.5649      R-Sq = 82.8%      R-Sq(adj) = 75.9%
```

Analysis of Variance

Source	DF	SS	MS	F	P
Regression	2	113142	56571	12.03	0.012
Residual Error	5	23506	4701		
Total	7	136648			

We can again use Minitab to predict y values for given values of the predictor variables. For example, to predict a bear's weight (y value), given that its head length (HEADLEN) is 14 in. and its LENGTH is 71.0 in., follow the same steps listed earlier in this section, but click on "Options" before clicking on the OK box. In the second dialog box, enter the values of 14.0 and 71.0 in the box identified as "Prediction intervals for new observations," then enter the desired confidence level, such as 95 (for 95% confidence). In addition to the elements in the above Minitab display, the following will also be shown:

Predicted Values for New Observations

New Obs	Fit	SE Fit	95.0% CI		95.0% PI	
1	306.3	43.9	(193.4,	419.2)	(97.0,	515.6)

This portion of the display shows that for a bear with a head length of 14.0 in. and a length of 71.0 in., the predicted weight is 306.3 lb. The prediction interval (for a single weight) is given, and the confidence interval (for the mean weight of all bears with 14.0 in. head length and 71.0 in. length) is also given.

When trying to find the best multiple regression equation, Minitab's **Stepwise Regression** procedure may be helpful. Select **Stat**, **Regression, Stepwise Regression**.

10-6 Modeling

Mathematical modeling is discussed in the Triola statistics textbooks (except *Essentials of Statistics*). The objective is to find a mathematical function that "fits" or describes real-world data. Among the models discussed in the textbook, we will describe how Minitab can be used for the linear, quadratic, logarithmic, exponential, and power models.

To illustrate the use of Minitab, consider the sample data in the table below. As in the textbook, we will use the coded year values for x, so that $x = 1, 2, 3, \ldots, 11$. The y values are the populations (in millions) of 5, 10, 17, ..., 281. For the following models, we enter 1, 2, 3, . .

. , 11 in column C1 and we enter 5, 10, 17, . . . , 281 in column C2. The resulting models are based on the coded values of x (1, 2, 3, . . . , 11) and the population values in millions (5, 10, 17, . . . , 281).

Population (in millions) of the United States

Year	1800	1820	1840	1860	1880	1900	1920	1940	1960	1980	2000
Coded year	1	2	3	4	5	6	7	8	9	10	11
Population	5	10	17	31	50	76	106	132	179	227	281

Linear Model: $y = a + bx$

The linear model can be obtained by using Minitab's correlation and regression module. The procedure is described in Section 10-3 of this manual/workbook. For the data in the above table, enter the coded year values of 1, 2, 3, . . . , 11 in column C1 and enter the population values of 5, 10, 17, . . . , 281 in column C2. The result will be $y = -61.9 + 2.72x$ with $R^2 = 92.5\%$ (or 0.925). The high value of R^2 suggests that the linear model is a reasonably good fit.

Quadratic Model: $y = ax^2 + bx + c$

With the coded values of x in column C1 and the population values (in millions) in column C2, select **Stat, Regression, Fitted Line Plot**. In the dialog box, enter C2 for the response variable, enter C1 for the predictor variable, and click on the **Quadratic** option. The results show that the function has the form given as $y = 10.01 - 6.003x + 2.767x^2$ with $R^2 = 0.999$. This higher value of R^2 suggests that the quadratic model is a better fit than the linear model.

Logarithmic Model: $y = a + b \ln x$

Minitab does not have a direct procedure for finding a logarithmic model, but there are ways to get it. One approach is to replace the coded x values (1, 2, 3, . . . , 11) with $\ln x$ values ($\ln 1$, $\ln 2$, $\ln 3$, . . . , $\ln 11$) This can be accomplished by using the command editor in the session window; enter the command LET C1 = LOGE(C1). [You could also use **Calc/Calculator** with the expression LOGE(C1).] After replacing the x values with $\ln x$ values, use **Stat, Regression, Fitted Line Plot** as in Section 10-3. That is, use the same procedure for finding the linear regression equation. The result is $y = -65.89 + 105.1 \ln x$, with $R^2 = 0.696$, suggesting that this model does not fit as well as the linear or quadratic models. Of the three models considered so far, the quadratic model appears to be best.

Exponential Model: $y = ab^x$

The exponential model is tricky, but it can be obtained using Minitab. Enter the values of x in the first column, and enter the values of $\ln y$ in the second column. The values of $\ln y$ can be found by using the command editor in the session window; enter the command LET C2 = LOGE(C2). [You could also use **Calc/Calculator** with the expression LOGE(C2).]

After replacing the y values with $\ln y$ values, use **Stat, Regression, Regression** to find the linear regression equation as described in Section 10-3. When you get the results from Minitab, the value of the coefficient of determination is correct, but the values of a and b in the exponential model must be computed as follows:

To find the value of a: Evaluate e^{b_0} where b_0 is the y-intercept given by Minitab.

To find the value of b: Evaluate e^{b_1} where b_1 is the slope given by Minitab.

Using the data in the above table, the value of $R^2 = 0.963$ is OK as is, but Minitab's regression equation of $y = 1.6556 + 0.39405x$ must be converted as follows.

$$a = e^{b_0} = e^{1.6556} = 5.2362$$
$$b = e^{b_1} = e^{0.39405} = 1.4830$$

Using these values of a and b, we express the exponential model as

$$y = 5.2362(1.4830^x)$$

Power Model: $y = ax^b$

Replace the x values with $\ln x$ values and replace the y values with $\ln y$ values. The values of $\ln x$ can be found by using the command editor in the session window; enter the command LET C1 = LOGE(C1). [You could also use **Calc/Calculator** with the expression LOGE(C1).] The values of $\ln y$ can be found the same way. Then use **Stat, Regression, Regression** to find the linear regression equation as described in Section 10-3 of this manual/workbook. When you get the results from Minitab, the value of the coefficient of determination is correct, the value of b is the same as the slope of the regression line, but the y-intercept must be converted as follows.

To find the value of a: Evaluate e^{b_0} where b_0 is the y-intercept given by Minitab.

Using the data in the above table, we get $R^2 = 0.976$. From Minitab's linear regression equation of $y = 1.2099 + 1.76606x$, we get $b = 1.76606$ and a is computed as follows.

$$a = e^{b_0} = e^{1.2099} = 3.35315$$

Using these values of a and b, we express the power model as

$$y = 3.35315(x^{1.76606})$$

The rationale underlying the methods for the exponential and power models is based on transformations of equations. In the exponential model of $y = ab^x$, for example, taking natural logarithms of both sides yields $\ln y = \ln a + x (\ln b)$, which is the equation of a straight line. Minitab can be used to find the equation of this straight line that fits the data best; the intercept will be $\ln a$ and the slope will be $\ln b$, but we need the values of a and b, so we solve for them as described above. Similar reasoning is used with the power model.

CHAPTER 10 EXPERIMENTS: Correlation and Regression

10–1. ***Bear Weights and Ages*** Refer to the Bears data set in Appendix B of the textbook. (The Minitab worksheet is BEARS.) Use the values for WEIGHT (x) and the values for AGE (y) to find the following.

    a.   Display the scatter diagram of the paired WEIGHT/AGE data. Based on that scatter diagram, does there appear to be a relationship between the weights of bears and their ages? If so, what is it?

              _____

    b.   Find the value of the linear correlation coefficient r. _____

    c.   Assuming a 0.05 level of significance, what do you conclude about the correlation between weights and ages of bears?

              _____

    d.   Find the equation of the regression line. (Use WEIGHT as the x predictor variable, and use AGE as the y response variable.) _____

    e.   What is the best predicted age of a bear that weighs 300 lb?_____

10–2. ***Effect of Transforming Data*** The ages used in Experiment 10–1 are in months. Convert them to days by multiplying each age by 30, then repeat Experiment 10–1 and enter the responses here:

    a.   Display the scatter diagram of the paired WEIGHT/AGE data. Based on that scatter diagram, does there appear to be a relationship between the weights of bears and their ages? If so, what is it?

              _____

    b.   Find the value of the linear correlation coefficient r._____

    c.   Assuming a 0.05 level of significance, what do you conclude about the correlation between weights and ages of bears?

              _____

    d.   Find the equation of the regression line. (Use WEIGHT as the x predictor variable, and use AGE as the y response variable.) _____

    e.   What is the best predicted age of a bear that weighs 300 lb?_____

    f.   After comparing the responses obtained in Experiment 10-1 to those obtained here, describe the general effect of changing the scale for one of the variables.

              _____

              _____

10–3. ***Bear Weights and Chest Sizes***   Refer to the Bears data set in Appendix B of the textbook. (The Minitab worksheet is BEARS.) Use the values for CHEST (x) and the values for WEIGHT (y) to find the following.

     a.   Display the scatter diagram of the paired CHEST/WEIGHT data. Based on that scatter diagram, does there appear to be a relationship between the chest sizes of bears and their weights? If so, what is it?

          _____

     b.   Find the value of the linear correlation coefficient r. _____

     c.   Assuming a 0.05 level of significance, what do you conclude about the correlation between chest sizes and weights of bears?

          _____

     d.   Find the equation of the regression line. (Use CHEST as the x predictor variable, and use WEIGHT as the y response variable.) _____

     e.   What is the best predicted weight of a bear with a chest size of 36.0 in? _____

     f.   When trying to obtain measurements from an anesthetized bear, what is a practical advantage of being able to predict the bear's weight by using its chest size?

          _____

          _____

10–4. ***Effect of No Variation for a Variable***   Use the following paired data and obtain the indicated results.

x	1	2	3	4	5	7	7	9
y	5	5	5	5	5	5	5	5

     a.   Print a scatter diagram of the paired x and y data. Based on the result, does there appear to be a relationship between x and y? If so, what is it?

          _____

     b.   What happens when you try to find the value of r? Why?

          _____

     c.   What do you conclude about the correlation between x and y? What is the equation of the regression line?

          _____

10–5. ***Buying a TV Audience*** The *New York Post* published the annual salaries (in millions) and the number of viewers (in millions), with results given below for Oprah Winfrey, David Letterman, Jay Leno, Kelsey Grammer, Barbara Walters, Dan Rather, James Gandolfini, and Susan Lucci, repsectively.

Salary	100	14	14	35.2	12	7	5	1
Viewers	7	4.4	5.9	1.6	10.4	9.6	8.9	4.2

Is there a correlation between salary and number of viewers? Explain.

_____

What is the equation of the regression line? _____

Find the best predicted value for the number of viewers (in millions), given that the salary (in millions of dollars) of television star Jennifer Anniston was $16 million. How does the predicted value compare to the actual number of viewers, which was 24 million?

_____

10–6. ***Supermodel Heights and Weights*** Listed below are heights (in inches) and weights (in pounds) for supermodels Niki Taylor, Nadia Avermann, Claudia Schiffer, Elle MacPherson, Christy Turlington, Bridget Hall, Kate Moss, Valerie Mazza, and Kristy Hume.

Height (in.)	71	70.5	71	72	70	70	66.5	70	71
Weight (lb.)	125	119	128	128	119	127	105	123	115

Is there a correlation between height and weight? If there is a correlation, does it mean that there is a correlation between height and weight of all adult women?

_____

What is the equation of the regression line? _____

Find the best predicted weight of a supermodel who is 69 in. tall. _____

10–7. ***Movie Data***   Refer to the data from movies in the table below, where all amounts are in millions of dollars. Let the movie gross amount be the dependent *y* variable.

Budget	62	90	50	35	200	100	90
Gross	65	64	48	57	601	146	47

Is there a correlation between the budget amount and the gross amount? Explain.

_____

(continued)

What is the equation of the regression line? _____

Find the best predicted gross amount for a movie with a budget of $150 million._____

10–8. *Crickets and Temperature* The numbers of chirps in one minute were recorded for different crickets, and the corresponding temperatures were also recorded. The results are given in the table below.

Chirps in one minute	882	1188	1104	864	1200	1032	960	900
Temperature (°F)	69.7	93.3	84.3	76.3	88.6	82.6	71.6	79.6

Is there a correlation between the number of chirps and the temperature? Explain.

_____

What is the equation of the regression line? _____

Find the best predicted temperature when a cricket chirps 1000 times in one minute.___

10–9. *Blood Pressure* Refer to the systolic and diastolic blood pressure measurements from randomly selected subjects. Let the dependent variable y represent diastolic blood pressure.

Systolic	138	130	135	140	120	125	120	130	130	144	143	140	130	150
Diastolic	82	91	100	100	80	90	80	80	80	98	105	85	70	100

Is there a correlation between systolic and diastolic blood pressure? Explain.

_____

What is the equation of the regression line? _____

Find the best predicted measurement of diastolic blood pressure for a person with a systolic blood pressure of 123. _____

10–10. *Garbage Data for Predicting Household Size* The "Garbage" data set in Appendix B of the textbook consists of data from the Garbage Project at the University of Arizona. (The Minitab worksheet is GARBAGE.) Use household size (HHSIZE) as the response y variable. For each predictor x variable given below, find the value of the linear correlation coefficient, the equation of the regression line, and the value of the coefficient of determination r^2. Enter the results in the spaces below.

(continued)

	_r_	Equation of regression line	_r_^{2}
Metal	___	_____	___
Paper	___	_____	___
Plastic	___	_____	___
Glass	___	_____	___
Food	___	_____	___
Yard	___	_____	___
Text	___	_____	___
Other	___	_____	___
Total	___	_____	___

Based on the above results, which single independent variable appears to be the best predictor of household size? Why?

_____

_____

10–11. Use the same data set described in Experiment 10-10. Let household size (HHSIZE) be the dependent _y_ variable and use the given predictor _x_ variables to fill in the results below.

	Multiple regression eq.	R^2	Adj. R^2
Metal and Paper	_____	___	_____
Plastic and Food	_____	___	_____
Metal, Paper, Glass	_____	___	_____
Metal, Paper, Plastic, Glass	_____	___	_____

Based on the above results, which of the multiple regression equations appears to best fit the data? Why?

_____

_____

10–12. ***Cigarette Data*** Use the "Cigarette Tar, Nicotine, and Carbon Monoxide" data set from Appendix B of the textbook. (The Minitab worksheet name is CIGARET.) Assume that we want to predict the amount of nicotine in a cigarette, based on the amount of tar and carbon monoxide. Use NICOTINE as the dependent variable and use TAR and/or CO (carbon monoxide) for independent variables. Find the equation that is best for predicting NICOTINE in a cigarette and describe why it is best.

_____

_____

10-13. ***Manatee Deaths from Boats*** Listed below are the numbers of Florida manatee deaths related to encounters with watercraft (based on data from *The New York Times*). The data are listed in order, beginning with the year 1980 and ending with the year 2000.

16 24 20 15 34 33 33 39 43 50 47 53 38 35 49 42 60 54 67 82 78

Use the given data to find equations and coefficients of determination for the indicated models.

	Equation	R^2
Linear	_____	_____
Quadratic	_____	_____
Logarithmic	_____	_____
Exponential	_____	_____
Power	_____	_____

Based on the above results, which model appears to best fit the data? Why?

_____

What is the best predicted value for 2001? In 2001, there were 82 watercraft–related manatee deaths. How does the predicted value compare to the actual value?

_____

11

Multinomial Experiments and Contingency Tables

11-1 Multinomial Experiments

Data from multinomial experiments consist of qualitative data that have been separated into different categories. The main objective is to determine whether the distribution agrees with or "fits" some claimed distribution. Also, a multinomial experiment is defined as follows:

Definition
A **multinomial experiment** is an experiment that meets the following conditions.

  1.   The number of trials is fixed.

  2.   The trials are independent.

  3.   All outcomes of each trial must be classified into exactly one
        of several different categories.

  4.   The probabilities for the different categories remain constant
        for each trial.

Minitab can now analyze data for a goodness-of-fit test. Use the following procedure.

Minitab Procedure for Multinomial Experiments

  1.   First collect the observed values *O* and enter them in column C1.

  2.   If the expected frequencies are not all the same, enter the expected *proportions*
        in column C2.

  3.   Select **Stat, Tables**, then select **Chi-Square Goodness-of-Fit Test**.

  4.   Enter C1 in the "Observed counts" box.

  5.   Select "Equal proportions" if the expected frequencies are all the same.
        Otherwise, select "Specific proportions" and enter C2 in the box.

  6.   Click **OK.**

The textbook includes an example involving the analysis of leading digits on the amounts of checks from companies suspected of fraud. The objective is to determine whether the leading digits fit the frequencies found from Benford's Law, whereby the leading digit of 1 occurs 30.1% of the time, 2 occurs 17.6% of the time, and so on, as in the table below.

Benford's Law: Distribution of Leadng Digits

Leading Digit	1	2	3	4	5	6	7	8	9
Frequency According to Benford's Law	30.1%	17.6%	12.5%	9.7%	7.9%	6.7%	5.8%	5.1%	4.6%
Expected Leading Digits of 784 Checks Following Benford's Law	235.984	137.984	98.000	76.048	61.936	52.528	45.472	39.984	36.064
Observed Leading Digits of 784 Checks Analyzed for Fraud	0	15	0	76	479	183	8	23	0

To use Minitab, enter the observed frequencies of 0, 15, 0, 76, 479, 183, 8, 23, and 0 in column C1, then enter these expected proportions in column C2: 0.301, 0.176, 0.125, 0.097, 0.079, 0.067, 0.058, 0.051, 0.046. After entering those values in columns C1 and C2, select **Stat, Tables,** select **Chi-Square Goodness-of-Fit Test** and use the dialog box as shown below.

The Minitab results are shown below.

Chi-Square Goodness-of-Fit Test for Observed Counts in Variable: C1

Category	Observed	Test Proportion	Expected	Contribution to Chi-Sq
1	0	0.301	235.984	235.98
2	15	0.176	137.984	109.61
3	0	0.125	98.000	98.00
4	76	0.097	76.048	0.00
5	479	0.079	61.936	2808.42
6	183	0.067	52.528	324.07
7	8	0.058	45.472	30.88
8	23	0.051	39.984	7.21
9	0	0.046	36.064	36.06

N	DF	**Chi-Sq**	**P-Value**
784	8	**3650.25**	**0.000**

The *P*-value of 0.000 suggests that we reject the null hypothesis that the frequencies fit the distribution determined by Benford's Law. There is a significant discrepancy between the distribution of the leading digits observed on the checks and the distribution that follows Benford's Law. This is a strong indication that the check amounts are not the result of typical transactions.

11-2 Contingency Tables

A **contingency table** (or **two-way frequency table**) is a table in which frequencies correspond to two variables. One variable is used to categorize rows, and a second variable is used to categorize columns. Let's consider the data in the contingency table shown below.

Case-Control Study of Motorcycle Drivers		
Color of Helmet		
Black	**White**	**Yellow/Orange**
Controls (not injured) 491	377	31
Cases (injured or killed) 213	112	8

Minitab can analyze data from a contingency table. To use Minitab with a contingency table, follow this procedure:

Minitab Procedure for Contingency Tables

1. Enter the individual columns of data in Minitab columns C1, C2, C3, . . . , .
For the data in the above table, enter 491 and 213 in column C1, then enter 377 and 112 in column C2, and so on. See the screen display shown below.

↓	C1	C2	C3
1	491	377	31
2	213	112	8

2. Select **Stat, Tables, Chisquare Test**.

3. Proceed to enter the names of the columns containing the observed frequencies. For the above data, enter C1 C2 C3. Click **OK**.

Shown below are the Minitab results from the above data. Important components have been highlighted in bold. We can see that the *P*-value is 0.012. If using a 0.05 significance level, we reject the null hypothesis of independence between the row and column variables. It appears that helmet color and group (control or case) are dependent. Because the controls were uninjured and the cases were injured or killed, it appears that there is an association between helmet color and motorcycle safety. The display also includes the test statistic of 8.775 as well as the expected value for each cell and the contribution of each cell to the total value of the chi-square test statistic.

Chi-Square Test: C1, C2, C3

```
Expected counts are printed below observed counts
Chi-Square contributions are printed below expected counts

            C1        C2       C3   Total
   1       491       377       31     899
        513.71    356.83    28.46
         1.004     1.140    0.227

   2       213       112        8     333
        190.29    132.17    10.54
         2.711     3.079    0.613

Total      704       489       39    1232
```

Chi-Sq = 8.775, DF = 2, **P-Value = 0.012**

11-3 Fisher's Exact Test

Fisher's exact test can be used for two-way tables. This text uses an *exact* distribution instead of an approximating chi-square distribution. It is particularly helpful when the approximating chi-square distribution cannot be used because of expected cell frequencies that are less than 5. Consider the sample data in the table below, with expected frequencies shown in parentheses. Note that the first cell has an expected frequency of 3, which is less than 5, so the chi-square distribution should not be used.

Helmets and Facial Injuries in Bicycle Accidents

	Helmet Worn	No Helmet
Facial injuries received	2 (3)	13 (12)
All injuries nonfacial	6 (5)	19 (20)

Minitab procedure for Fisher's Exact Test

1.   Enter the frequencies in column C1 with the row numbers in column C2 and the   column numbers in column C3. Here are the Minitab entries for the above table. (The   column   labels of Frequency, Row, and Column were manually entered.)

↓	C1 Frequency	C2 Row	C3 Column
1	2	1	1
2	13	1	2
3	6	2	1
4	19	2	2

2.   Select **Stat, Tables,** then **Cross Tabulation and Chi-Square.**

3.   Enter C2 in the "For rows" box, enter C3 in the "For columns" box, and enter C1 in the "Frequencies are in" box.

4.   Click on the **Other Stats** bar and select "Fisher's exact test for 2 × 2 tables."

5.   Click **OK** twice to get the resulting *P*-value. Here is the display that results from the above table: **Fisher's exact test: P-Value = 0.685661.** Because the *P*-value is large, we fail to reject the null hypothesis that wearing a helmet and receiving facial injuries are independent.

CHAPTER 11:   Multinomial Experiments and Contingency Tables

11-1. *Loaded Die* The author drilled a hole in a die and filled it with a lead weight, then proceeded to roll it 200 times. Here are the observed frequencies for the outcomes of 1, 2, 3, 4, 5, and 6 respectively: 27, 31, 42, 40, 28, 32. Use a 0.05 significance level to test the claim that the outcomes are not equally likely.

Test statistic:_____   *P*-value:_____

Conclusion:_____

_____

Does it appear that the loaded die behaves differently than a fair die?

_____

11–2. *Flat Tire and Missed Class* A classic tale involves four car-pooling students who missed a test and gave as an excuse a flat tire. On the makeup test, the instructor asked the students to identify the particular tire that went flat. If they really didn't have a flat tire, would they be able to identify the same tire? The author asked 41 other students to identify the tire they would select. The results are listed in the following table (except for one student who selected the spare). Use a 0.05 significance level to test the author's claim that the results fit a uniform distribution.

Tire	Left front	Right front	Left rear	Right rear
Number selected	11	15	8	6

Test statistic:_____   *P*-value:_____

Conclusion:_____

_____

What does the result suggest about the ability of the four students to select the same tire when they really didn't have a flat?

_____

11–3. *Grade and Seating Location* Do "A" students tend to sit in a particular part of the classroom? The author recorded the locations of the students who received grades of A, with these results: 17 sat in the front, 9 sat in the middle, and 5 sat in the back of the classroom. Is there sufficient evidence to support the claim that the "A" students are not evenly distributed throughout the classroom? If so, does that mean you can increase your likelihood of getting an A by sitting in the front?

Test statistic:_____   *P*-value:_____

Conclusion:_____

_____

11–4. ***Measuring Pulse Rates*** According to one procedure used for analyzing data, when certain quantities are measured, the last digits tend to be uniformly distributed, but if they are estimated or reported, the last digits tend to have disproportionately more 0s or 5s. Refer to Data Set 1 in Appendix B of the textbook and use the last digits of the pulse rates of the 80 men and women. Those pulse rates were obtained as part of the National Health and Examination Survey. Test the claim that the last digits of 0, 1, 2, 3, . . . , 9 occur with the same frequency.

Test statistic:_____   *P*-value:_____

Conclusion:_____

_____

Based on the observed digits, what can be inferred about the procedure used to obtain the pulse rates?

_____

11–5. ***Genetics Experiment*** Based on the genotypes of parents, offspring are expected to have genotypes distributed in such a way that 25% have genotypes denoted by AA, 50% have genotypes denoted by Aa, and 25 % have genotypes denoted by aa. When 145 offspring are obtained, it is found that 20 of them have AA genotypes, 90 have Aa genotypes and 35 have aa genotypes. Test the claim that the observed genotype offspring frequencies fit the expected distribution of 25% for AA, 50% for Aa, and 25% for aa. Use a significance level of 0.05.

Test statistic:_____   *P*-value:_____

Conclusion:_____

_____

11-6. ***Testing a Normal Distribution***   In this experiment we will use Minitab's ability to generate normally distributed random numbers. We will then test the sample data to determine if they actually do fit a normal distribution.

a. Generate 1000 random numbers from a normal distribution with a mean of 100 and a standard deviation of 15. (IQ scores have these parameters.) Select **Calc**, then **Random Data**, then **Normal**.

b. Use **Data/Sort** (or **Manip/Sort** on Minitab Release 13 and earlier) to arrange the data in order.

c. Examine the sorted list and determine the frequency for each of the categories listed below. Enter those frequencies in the spaces provided. (The expected frequencies are found by assuming that the data are normally distributed.)

	Observed Frequency	Expected Frequency
Below 55:	_____	1
55-70:	_____	22
70-85:	_____	136
85-100:	_____	341
100-115:	_____	341
115-130:	_____	136
130-145:	_____	22
Above 145:	_____	1

d. Use Minitab to test the claim that the randomly generated numbers actually do fit a normal distribution with mean 100 and standard deviation 15.

Test statistic:_____   *P*-value:_____

Conclusion:_____

_____

11-7. **E–Mail and Privacy** Workers and senior–level bosses were asked if it was seriously unethical to monitor employee e–mail, and the results are summarized in the table (based on data from a Gallup poll). Use a 0.05 significance level to test the claim that the response is independent of whether the subject is a worker or a senior–level boss.

	Yes	No
Workers	192	244
Bosses	40	81

Test statistic:_____   *P*-value:_____

Conclusion:_____

_____

Does the conclusion change if a significance level of 0.01 is used instead of 0.05? Do workers and bosses appear to agree on this issue?

_____

11–8. **Accuracy of Polygraph Tests** The data in the accompanying table summarize results from tests of the accuracy of polygraphs (based on data from the Office of Technology Assessment). Use a 0.05 significance level to test the claim that whether the subject lies is independent of the polygraph indication.

	Polygraph Indicated Truth	Polygraph Indicated Lie
Subject actually told the truth	65	15
Subject actually told a lie	3	17

Test statistic:_____   *P*-value:_____

Conclusion:_____

_____

What do the results suggest about the effectiveness of polygraphs?

_____

11–9. *Fear of Flying Gender Gap*   The Marist Institute for Public Opinion conducted a poll of 1014 adults, 48% of whom were men. The poll results show that 12% of the men and 33% of the women fear flying. After constructing a contingency table that summarizes the data in the form of frequency counts, use a 0.05 significance level to test the claim that gender is independent of the fear of flying.

Test statistic:_____   *P*-value:_____

Conclusion:_____

_____

11–10. *Occupational Hazards* Use the data in the table to test the claim that occupation is independent of whether the cause of death was homicide. The table is based on data from the U.S. Department of Labor, Bureau of Labor Statistics.

	Police	Cashiers	Taxi Drivers	Guards
Homicide	82	107	70	59
Cause of Death Other than Homicide	92	9	29	42

Test statistic:_____   *P*-value:_____

Conclusion:_____

_____

Does any particular occupation appear to be most prone to homicides? If so, which one?

_____

How are the results affected if the order of the rows is switched?

_____

How are the results affected by the presence of an outlier? If we change the first entry from 82 to 8200, are the results dramatically affected?

_____

11-11. ***Fisher's Exact Test*** Refer to Experiment 11-7 in this manual/workbook. Repeat that experiment by using Fisher's exact test instead of using the approximating chi-square distribution. Enter the results below.

P-value obtained by using the approximating chi-square distribution: _____

P-value obtained by using Fisher's exact test: _____

Does the use of the Fisher's exact test have much of an effect on the *P*-value?

_____

11-12. ***Fisher's Exact Test*** Refer to Experiment 11-8 in this manual/workbook. Repeat that experiment by using Fisher's exact test instead of using the approximating chi-square distribution. Enter the results below.

P-value obtained by using the approximating chi-square distribution: _____

P-value obtained by using Fisher's exact test: _____

Does the use of the Fisher's exact test have much of an effect on the *P*-value?

_____

11-13. ***Fisher's Exact Test*** Refer to Experiment 11-9 in this manual/workbook. Repeat that experiment by using Fisher's exact test instead of using the approximating chi-square distribution. Enter the results below.

P-value obtained by using the approximating chi-square distribution: _____

P-value obtained by using Fisher's exact test: _____

Does the use of the Fisher's exact test have much of an effect on the *P*-value?

_____

12

Analysis of Variance

12-1 One-Way Analysis of Variance

One–way analysis of variance is used to test the claim that three or more populations have the same mean. When the textbook discusses one-way analysis of variance, it is noted that the term "one-way" is used because the sample data are separated into groups according to one characteristic or "factor". In the table below, the factor is the type of treatment. That is, the weights of the poplar trees are analyzed in the context of the factor consisting of four separate treatment categories.

Samples of the same size are considered along with samples of different sizes. Minitab will work with both types; the samples need not have the same number of values.

Weights (kg) of Poplar Trees

	Treatment			
	None	Fertilizer	Irrigation	Fertilizer and Irrigation
	0.15	1.34	0.23	2.03
	0.02	0.14	0.04	0.27
	0.16	0.02	0.34	0.92
	0.37	0.08	0.16	1.07
	0.22	0.08	0.05	2.38
n	5	5	5	5
$\overline{x}$	0.184	0.332	0.164	1.334
s	0.127	0.565	0.126	0.859

The following procedure describes how Minitab can be used with such collections of sample data to test the claim that the different samples come from populations with the same mean. For the data in the above table, the claim of equal means leads to these hypotheses:

H_0: $\mu_1 = \mu_2 = \mu_3 = \mu_4$
H_1: At least one of the population means is different from the others.

Minitab Procedure for One-Way Analysis of Variance

1. Enter the original sample values in columns.

    (*Note:* For each set of sample values, if you know only the summary statistics of n, $\overline{x}$, and s, but the original sample values are not available, you must create a list of n values having the same mean $\overline{x}$ and the same standard deviation s. See Section 6–5 of this manual/workbook.)

2. Select **Stat** from the main menu.

3. Select the subdirectory item of **ANOVA**.

4.   Select the option of **Oneway (Unstacked)**. The term "unstacked" means that the data are not all stacked in one single column; they are listed in their individual and separate columns.

5.   In the dialog box, enter the column names (C1 C2 C3 C4) in the box labeled as "Responses (in separate columns)." Click **OK**.

If you use the above steps with the weights from the above table, the Minitab results will appear as shown below.

One-way ANOVA: None, Fert, Irrig, Fert&Irrig

```
Source   DF      SS      MS      F       P
Factor    3   4.682   1.561   5.73   0.007
Error    16   4.357   0.272
Total    19   9.040

S = 0.5219   R-Sq = 51.80%   R-Sq(adj) = 42.76%


                                Individual 95% CIs For Mean Based on
                                Pooled StDev
Level         N    Mean   StDev  ------+---------+---------+---------+--
None          5  0.1840  0.1270   (--------*-------)
Fert          5  0.3320  0.5651    (--------*-------)
Irrig         5  0.1640  0.1262  (--------*-------)
Fert&Irrig    5  1.3340  0.8590                     (-------*-------)
                                 ------+---------+---------+---------+--
                                    0.00      0.60      1.20      1.80
Pooled StDev = 0.5219
```

The *P*-value of 0.007 indicates that there is sufficient sample evidence to warrant rejection of the null hypothesis that $\mu_1 = \mu_2 = \mu_3 = \mu_4$. The test statistic of $F = 5.73$ is also provided, as are the values of the SS and MS components. The lower portion of the Minitab display also includes the individual sample means, standard deviations, and graphs of the 95% confidence interval estimates of each of the four population means. These confidence intervals are computed by the same methods used in Chapter 7, except that a pooled standard deviation is used instead of the individual sample standard deviations. (The last entry in the Minitab display shows that the pooled standard deviation is 0.5219.)

Note that the Minitab display uses the term *factor* instead of *treatment*, so that the value of SS(treatment) is 4.682, and MS(treatment) = 1.561.

Caution: It is easy to feed Minitab data that can be processed quickly and painlessly, but we should *think* about what we are doing. We should consider the assumptions for the test being used, and we should *explore* the data before jumping into a formal procedure such as analysis of variance. Carefully explore the important characteristics of data, including the center (through means and medians), variation (through standard deviations and ranges), distribution (through histograms and boxplots), outliers, and any changing patterns over time.

12-2 Two-Way Analysis of Variance

Two-way analysis of variance involves *two* factors, such as treatment and site in the table below. The two–way analysis of variance procedure requires that we test for (1) an interaction effect between the two factors; (2) an effect from the row factor; (3) an effect from the column factor.

Poplar Tree Weights (kg)

	No Treatment	Fertilizer	Irrigation	Fertilizer and Irrigation
Site 1 **(rich, moist)**	0.15 0.02 0.16 0.37 0.22	1.34 0.14 0.02 0.08 0.08	0.23 0.04 0.34 0.16 0.05	2.03 0.27 0.92 1.07 2.38
Site 2 **(sandy, dry)**	0.60 1.11 0.07 0.07 0.44	1.16 0.93 0.30 0.59 0.17	0.65 0.08 0.62 0.01 0.03	0.22 2.13 2.33 1.74 0.12

The first obstacle to overcome is to understand a somewhat awkward method of entering sample data. When entering the values in the above table, we must somehow keep track of the location of each data value, so here is the procedure to be used:

1.   Enter all of the sample data in column C1.

2.   Enter the corresponding row numbers in column C2.

3.   Enter the corresponding column numbers in column C3.

This particular format can be confusing at first, so stop and try to recognize the pattern. Because the first value of 0.15 from the above table is in the first row and first column, its row identification is 1 and its column identification is 1. The entry of 1.34 is in the first row and second column, so its row identification number is 1 and its column identification number is 2. Also, its wise to name column C1 as "WEIGHT," name column C2 as "SITE," and name column C3 as "Treatment" so that the columns have meaningful names. Here is an illustration of the pattern used to enter the sample data in the above table.

```
C1          C2          C3
WEIGHT      SITE        TREATMENT
0.15        1           1              ←Row 1 and column 1
0.02        1           1              ←Row 1 and column 1
  .           .           .
  .           .           .
  .           .           .
0.12        2           4              ←Row 2 and column 4
```

The procedure for obtaining a Minitab display for two-way analysis of variance is as follows.

1.   Enter all of the sample values in column C1 (as shown above).

2.   Enter the corresponding row numbers in column C2 (as shown above).

3.   Enter the corresponding column numbers in column C3 (as shown above).

4.   Select **Stat**, then **ANOVA**, then **Twoway**.

5.   You should now see a dialog box like the one shown on the next page.

     Make these entries in the dialog box:

     • Enter C1 in the box labeled Response.

     • Enter C2 in the box labeled Row Factor.

     • Enter C3 in the box labeled Column Factor.

     • Click **OK**.

Using the data in the preceding table will result in the Minitab display shown below.

Two-way ANOVA: WEIGHT versus SITE, TREATMENT

```
Source        DF      SS       MS       F      P
SITE           1   0.2722  0.27225   0.81  0.374
TREATMENT      3   7.5470  2.51567   7.50  0.001
Interaction    3   0.1716  0.05721   0.17  0.915
Error         32  10.7267  0.33521
Total         39  18.7176
```

$$S = 0.5790 \quad R\text{-Sq} = 42.69\% \quad R\text{-Sq(adj)} = 30.16\%$$

The Triola textbooks (excluding *Essentials of Statistics*) describe the interpretation of the preceding Minitab display.

1. Test for Interaction
We begin by testing the null hypothesis that there is no *interaction* between the two factors of SITE and TREATMENT. Using the above Minitab results, we calculate the following test statistic.

$$F = \frac{MS(interaction)}{MS(error)} = \frac{0.05721}{0.33521} = 0.17$$

Minitab shows the corresponding *P*-value of 0.915, so we fail to reject the null hypothesis of no interaction between the two factors of SITE and TREATMENT. There does not appear to be an effect from an interaction between site and treatment.

2. Test for Effect from the Row Factor

Our two-way analysis of variance procedure outlined in the textbook indicates that we should now proceed to test these two null hypotheses: (1) There are no effects from the row (SITE) factor; (2) There are no effects from the column (TREATMENT) factor. For the test of an effect from the row factor (site), we have

$$F = \frac{MS(SITE)}{MS(error)} = \frac{0.27225}{0.33521} = 0.81$$

The Minitab *P*-value is 0.374, so we fail to reject the claim that there is no effect from site. That is, the site of the trees does not appear to have an effect on their weight.

3. Test for Effect from the Column Factor

For the test of an effect from treatment, we have

$$F = \frac{MS(TREATMENT)}{MS(error)} = \frac{2.51567}{0.33521} = 7.50$$

The Minitab *P*–value is displayed as 0.001, so treatment does appear to have an effect on weight.

Special Case: One Observation Per Cell

The Triola textbook (excluding *Essentials of Statistics*) includes a subsection describing the special case in which there is only one sample value in each cell. Here's how we proceed when there is one observation per cell: *If it seems reasonable to assume (based on knowledge about the circumstances) that there is no interaction between the two factors, make that assumption and then proceed as before to test the following two hypotheses separately:*

H_0: There are no effects from the row factor.

H_0: There are no effects from the column factor.

To use Minitab, simply apply the same procedure described earlier. The Minitab results will not include the values for an interaction, but the other necessary values are provided.

CHAPTER 12 EXPERIMENTS: Analysis of Variance

In Experiments 12–1 and 12–2, use the listed sample data from car crash experiments conducted by the National Transportation Safety Administration. New cars were purchased and crashed into a fixed barrier at 35 mi/h, and the listed measurements were recorded for the dummy in the driver's seat. The subcompact cars are the Ford Escort, Honda Civic, Hyundai Accent, Nissan Sentra, and Saturn SL4. The compact cars are Chevrolet Cavalier, Dodge Neon, Mazda 626 DX, Pontiac Sunfire, and Subaru Legacy. The Midsize cars are Chevrolet Camaro, Dodge Intrepid, Ford Mustang, Honda Accord, and Volvo S70. The full–size cars are Audi A8, Cadillac Deville, Ford Crown Victoria, Oldsmobile Aurora, and Pontiac Bonneville.

12–1. **Head Injury in a Car Crash**   The head injury data (in hic) are given below. Use a 0.05 significance level to test the null hypothesis that the different weight categories have the same mean. Do the data suggest that larger cars are safer?

Subcompact:	681	428	917	898	420
Compact:	643	655	442	514	525
Midsize:	469	727	525	454	259
Full–Size:	384	656	602	687	360

SS(treatment): _____   MS(treatment): _____   Test statistic F: _____

SS(error): _____   MS(error): _____ P-value: _____

SS(total): _____

Conclusion:_____

12–2. **Chest Deceleration in a Car Crash**   The chest deceleration data (g) are given below. Use a 0.05 significance level to test the null hypothesis that the different weight categories have the same mean. Do the data suggest that larger cars are safer?

Subcompact:	55	47	59	49	42
Compact:	57	57	46	54	51
Midsize:	45	53	49	51	46
Full–Size:	44	45	39	58	44

SS(treatment): _____   MS(treatment): _____   Test statistic F: _____

SS(error): _____   MS(error): _____ P-value: _____

SS(total): _____

Conclusion:_____

12–3. *Archeology: Skull Breadths from Different Epochs*   The values in the table are measured maximum breadths of male Egyptian skulls from different epochs (based on data from Ancient Races of the Thebaid, by Thomson and Randall-Maciver). Changes in head shape over time suggest that interbreeding occurred with immigrant populations. Use a 0.05 significance level to test the claim that the different epochs do not all have the same mean.

4000 B.C.	1850 B.C.	150 A.D.
131	129	128
138	134	138
125	136	136
129	137	139
132	137	141
135	129	142
132	136	137
134	138	145
138	134	137

SS(treatment): _____   MS(treatment): _____   Test statistic F: _____

SS(error): _____   MS(error): _____ P-value: _____

SS(total): _____

Conclusion: _____

_____

12–4. *Mean Weights of M&Ms*   Refer to the M&M data set in Appendix B from the textbook. At the 0.05 significance level, test the claim that the mean weight of M&Ms is the same for each of the six different color populations.

SS(treatment): _____   MS(treatment): _____   Test statistic F: _____

SS(error): _____   MS(error): _____ P-value: _____

SS(total): _____

Conclusion: _____

If it is the intent of Mars, Inc., to make the candies so that the different color populations have the same mean weight, do these results suggest that the company has a problem requiring corrective action?

12–5. *Homerun Distances*   Refer to the HOMERUNS data set in Appendix B from the textbook. Use a 0.05 significance level to test the claim that the homeruns hit by Barry Bonds, Mark McGwire, and Sammy Sosa have mean distances that are not all the same.

SS(treatment): _____   MS(treatment): _____   Test statistic F: _____

SS(error): _____   MS(error): _____ P-value: _____

SS(total): _____

Conclusion: _____

_____

Do the homerun distances explain the fact that as of this writing, Barry Bonds has the most homeruns in one season, while Mark McGwire has the second highest number of runs?

_____

12–6. *Exercise and Stress*   A study was conducted to investigate the effects of exercise on stress. The table below lists systolic blood pressure readings (in mm Hg) of subjects from the time preceding 25 minutes of aerobic bicycle exercise and preceding the introduction of stress through arithmetic and speech tests (based on data from "Sympathoadrenergic Mechanisms in Reduced Hemodynamic Stress Responses after Exercise" by Kim Brownley et al, *Medicine and Science in Sports and Exercise*, Vol. 35, No. 6). Use a 0.05 significance level to test the claim that the different groups of subjects have the same mean blood pressure. Enter the results in the spaces that follow the table.

Female/Black	Male/Black	Female/White	Male/White
117.00	115.67	119.67	124.33
130.67	120.67	106.00	111.00
102.67	133.00	108.33	99.67
93.67	120.33	107.33	128.33
96.33	124.67	117.00	102.00
92.00	118.33	113.33	127.33

SS(treatment): _____   MS(treatment): _____   Test statistic F: _____

SS(error): _____   MS(error): _____ P-value: _____

SS(total): _____

Conclusion: _____

12-7.   ***Simulations***   Use Minitab to randomly generate three different samples of 500 values each. (Select **Calc, Random Data, Normal**.) For the first two samples, use a normal distribution with a mean of 100 and a standard deviation of 15. For the third sample, use a normal distribution with a mean of 101 and a standard deviation of 15. We know that the three populations have different means, but do the methods of analysis of variance allow you to conclude that the means are different? Explain.

12–8.   ***SAT Scores***   The sample data in the table are SAT scores on the verbal and math portions of SAT-I and are based on reported statistics from the College Board.

Verbal

| Female | 646 | 539 | 348 | 623 | 478 | 429 | 298 | 782 | 626 | 533 |
| Male | 562 | 525 | 512 | 576 | 570 | 480 | 571 | 555 | 519 | 596 |

Math

| Female | 484 | 489 | 436 | 396 | 545 | 504 | 574 | 352 | 365 | 350 |
| Male | 547 | 678 | 464 | 651 | 645 | 673 | 624 | 624 | 328 | 548 |

Test the null hypothesis that SAT scores are not affected by an interaction between gender and test (verbal/math).

Assume that SAT scores are not affected by an interaction between gender and the type of test (verbal/math). Is there sufficient evidence to support the claim that gender has an effect on SAT scores?

Assume that SAT scores are not affected by an interaction between gender and the type of test (verbal/math). Is there sufficient evidence to support the claim that the type of test (verbal/math) has an effect on SAT scores?

12–9. ***Pulse Rates*** Use the following pulse rates from Data Set 1 in Appendix B.

	Age		
	Under 20	20–40	Over 40
Male	96 64 68 60	64 88 72 64	68 72 60 88
Female	76 64 76 68	72 88 72 68	60 68 72 64

Are pulse rates affected by an interaction between gender and age? Explain.

_____

Are pulse rates affected by gender? Explain.

_____

Are pulse rates affected by age? Explain.

_____

12–10. ***Estimating Distance*** The sample data in the table are student estimates (in feet) of the length of their classroom. The actual length of the classroom is 24 ft 7.5 in.

	Major		
	Math	Business	Liberal Arts
Female	28 25 30	35 25 20	40 21 30
Male	25 30 20	30 24 25	25 20 32

What do you conclude about an interaction between gender and major?

_____

Assume that the estimated distances are not affected by an interaction between gender and major. Do the estimated distances appear to be affected by gender? Explain.

_____

Assume that the estimated distances are not affected by an interaction between gender and major. Do the estimated distances appear to be affected by major? Explain.

_____

12–11. ***Estimating Distance*** Refer to Experiment 12-10 and use only the first value in each cell. Assume that the estimated distances are not affected by an interaction between gender and major.

Do the estimated distances appear to be affected by gender? Explain.

_____

Do the estimated distances appear to be affected by major? Explain.

_____

13

Nonparametric Statistics

13-1 Ranking Data

This chapter describes how Minitab can be used for the nonparametric methods presented in Chapter 13 of *Elementary Statistics,* 10th edition. The sections of this chapter correspond to those in the textbook. The textbook introduces some basic principles of nonparametric methods, but it also describes a procedure for converting data into their corresponding *ranks*. Here is the Minitab procedure for using data in column C1 to create a column C2 that consists of the corresponding ranks.

Converting Data to Ranks

    1. Enter the data values in column C1.

    2. Select **Data** from the main menu. (For Minitab Release 13 and earlier, select **Manip** from the main menu.)

    3. Select **Rank** from the subdirectory.

    4. In the dialog box, make these entries:

        • For the box labeled "Rank data in," enter C1.

        • For the box labeled "Store ranks in," enter C2.

        • Click **OK**.

    As an example, consider the values of 5, 3, 40, 10, 12, and 12. They are converted to the ranks of 2, 1, 6, 3, 4.5, and 4.5. Enter the values of 5, 3, 40, 10, 12, and 12 in column C1, follow the above steps, and column C2 will contain the ranks of 2, 1, 6, 3, 4.5, and 4.5 as shown below.

↓	C1	C2
1	5	2.0
2	3	1.0
3	40	6.0
4	10	3.0
5	12	4.5
6	12	4.5

Ties: Note that Minitab handles ties as described in the textbook. See the above display that includes ties in the ranks that correspond to the values of 12 and 12. Ties occur with the corresponding ranks of 4 and 5, so Minitab assigns a rank of 4.5 to each of those values.

13-2 Sign Test
The Triola textbook (excluding *Essentials of Statistics*) includes the following definition.

Definition
The **sign test** is a nonparametric (distribution-free) test that uses plus and minus signs to test different claims, including these:

1. Given matched pairs of sample data, test claims about the medians of the two populations.
2. Given nominal data, test claims about the proportion of some category.
3. Test claims about the median of a single population.

Minitab makes it possible to work with all three of the above cases. Let's consider the matched data in the table below. (The data are matched, because each pair of values is from *adjacent* plots of land.)

Yields of Corn from Different Seeds

Regular	1903	1935	1910	2496	2108	1961	2060	1444	1612	1316	1511
Kiln Dried	2009	1915	2011	2463	2180	1925	2122	1482	1542	1443	1535
Sign of difference	−	+	−	+	−	+	−	−	+	−	−

Given such paired data, we can use Minitab to apply the sign test to test the claim of no difference.

Minitab Procedure for Sign Test with Matched Data

1. Enter the values of the first variable in column C1.

2. Enter the values of the second variable in column C2.

3. Create column C3 equal to the values of C1 − C2. Accomplish this using either of the following two approaches.

   Session window: Click on the session window at the top portion of the screen, then click on **Editor**, then click on **Enable Commands**. Enter the command:
   LET C3 = C1 - C2.

   Calculator: Click on **Calc**, then **Calculator**. In Minitab's calculator, enter C3 for the "Store result in variable" box, and enter the expression C1 - C2 in the "expression" box. Click **OK** when done.

4. Select **Stat** from the main menu.

5. Select **Nonparametrics** from the subdirectory.

6. Select the option of **1-Sample Sign**.

7. Make these entries in the resulting dialog box:

- For the box labeled Variables, enter C3 (which contains the differences obtained when column C2 is subtracted from column C1).

- Click on the small circle for **Test median** and, in the adjacent box, enter 0 for a test of the claim that the median is equal to zero.

- In the box labeled Alternative, select **not equal,** so that the alternative hypothesis is that the statement that the median is not equal to zero.

- Click **Ok**.

Here are the Minitab results for the data in the above table:

```
Sign test of median =  0.00000 versus not = 0.00000

      N   Below   Equal   Above       P   Median
C3   11       7       0       4  0.5488   -38.00
```

The above results show that there are 11 pairs of data, there are 7 differences that are below zero (negative sign), none of the differences is equal to zero, and 4 differences are above zero (positive sign). The P-value is 0.5488. In the textbook, we would use a decision criterion that involves a comparison of the test statistic $x = 4$ (the less frequent sign) and a critical value found in Table A-7. Because Minitab displays the P-value, we can use the P-value approach to hypothesis testing. Because the P-value of 0.5488 is greater than the significance level of $\alpha = 0.05$, there is not sufficient evidence to warrant rejection of the claim that the median of the differences is equal to 0. There does not appear to be significant difference between the yields from regular seeds and kiln-dried seeds.

Sign test with nominal data: In some cases, the data for the sign test do not consist of values of the type given in the table above. For example, suppose that a test of a gender-selection method results in 325 babies, including 295 girls. Because the data are at the nominal level of measurement, we could enter them as −1's (for girls) and 1's (for boys). *Shortcut*: Instead of entering −1 a total of 295 times and 1 a total of 30 times, we can enter the following commands in the session window to achieve the same result. (Click on the session window at the top portion of the screen, then click on **Editor**, then click on **Enable Commands**. Enter these commands to create a column consisting of −1 repeated 295 times and 1 repeated 30 times.)

```
SET C3
295(-1), 30(1)
END
```

Now follow the above steps (beginning with step 3) for conducting the sign test, and the results will be as shown below.

Sign test of median = 0.00000 versus not = 0.00000

	N	Below	Equal	Above	P	Median
C3	325	295	0	30	**0.0000**	-1.000

The resulting *P*-value of 0.0000 suggests that we reject the null hypothesis of equality of girls and boys. There is sufficient evidence to warrant rejection of the claim that girls and boys are equally likely.

Sign test for testing a claimed value of the median: The textbook also includes an example involving body temperature data. (See the body temperatures for 12:00 AM on day 2.) We can apply the sign test to test the claim that the *median* is less than 98.6°. Enter the 106 body temperatures in column C3 and follow the steps for doing the sign test, but make these changes in the dialog box entries of Step 6: Enter 98.6 in the box adjacent to "Test median," and select "less than" in the box labeled Alternative. The Minitab results will be as shown below.

Sign test of median = 98.60 versus < 98.60

	N	Below	Equal	Above	P	Median
C1	106	68	15	23	**0.0000**	98.40

The above Minitab display shows that among the 106 body temperatures, 68 are below 98.6, 15 are equal to 98.6, and 23 are above 98.6. The *P*-value of 0.0000 causes us to reject the null hypothesis that the median is at least 98.6. There appears to be sufficient evidence to support the claim that the median body temperature is less than 98.6° F.

13-3 Wilcoxon Signed-Ranks Test

The Triola textbook (excluding *Essentials of Statistics*) describes the Wilcoxon signed-ranks test, and the following definition is given.

Definition
The **Wilcoxon signed-ranks test** is a nonparametric test that uses ranks of sample data consisting of *matched pairs*. It is used to test the null hypothesis that the population of differences has a median of zero, so the null and alternative hypotheses are as follows:

H_0: The matched pairs have differences that come from a population with a median equal to zero.

H_1: The matched pairs have differences that come from a population with a nonzero median.

The textbook makes this important point: The Wilcoxon signed-ranks test and the sign test can both be used with sample data consisting of matched pairs, but the sign test uses only the signs of the differences and not their actual magnitudes (how large the numbers are). The Wilcoxon signed-ranks test uses ranks, so the magnitudes of the differences are taken into account. Because the Wilcoxon signed-ranks test incorporates and uses more information than the sign test, it tends to yield conclusions that better reflect the true nature of the data. First we describe the Minitab procedure for conducting a Wilcoxon signed-ranks test, then we illustrate it with an example.

1. Enter the values of the first variable in column C1.

2. Enter the values of the second variable in column C2.

3. Create column C3 equal to the values of C1 − C2. Accomplish this using either of the following two approaches.

   Session window: Click on the session window at the top portion of the screen, then click on **Editor**, then click on **Enable Commands**. Enter the command:
   LET C3 = C1 − C2.

   Calculator: Click on **Calc**, then **Calculator**. In Minitab's calculator, enter C3 for the "Store result in variable" box, and enter the expression C1 − C2 in the "expression" box. Click **OK** when done.

4. Select **Stat** from the main menu.

5. Select **Nonparametrics** from the subdirectory.

6. Select the option of **1-Sample Wilcoxon**.

7. Make these entries in the resulting dialog box:

- For the box labeled Variables, enter C3 (which contains the differences obtained when column C2 is subtracted from column C1).

- Click on the small circle for "Test median," and, in the adjacent box, enter 0 for a test of the claim that the median is equal to zero.

- In the box labeled Alternative, select "not equal" for a test of the claim that the median is equal to zero (so that the alternative hypothesis is that the median is not equal to zero).

- Click **OK**.

Using the same matched data from the table in Section 13-2 of this manual/workbook, the Minitab results for the Wilcoxon signed-ranks test will be as shown below.

Wilcoxon Signed Rank Test: C3
```
Test of median = 0.000000 versus median not = 0.000000
             N
           for   Wilcoxon              Estimated
     N    Test   Statistic       P      Median
C3   11    11        15.0    0.120     -34.50
```

From these results we see that the test statistic is 15.0 and the P-value is 0.120. Because the P-value of 0.120 is greater than the significance level of 0.05, we fail to reject the null hypothesis. There does not appear to be a significant difference between the regular seeds and the kiln-dried seeds.

13-4 Wilcoxon Rank-Sum Test

The Triola textbook (excluding *Essentials of Statistics*) discusses the Wilcoxon rank-sum test and includes the following definition.

Definition
The **Wilcoxon rank-sum test** is a nonparametric test that uses ranks of sample data from two *independent* populations. It is used to test the null hypothesis that the two independent samples come from populations with the same distribution. (That is, the two populations are identical.) The alternative hypothesis is the claim that the two population distributions are different in some way.

H_0:   The two samples come from populations with equal medians.
H_1:   The two samples come from populations with different medians.

The Wilcoxon rank-sum test described in the textbook is equivalent to the Mann-Whitney U test, so use this Minitab Mann-Whitney procedure for testing the claim that two independent samples come from populations with populations with equal medians:

1. Enter the two sets of sample data in columns C1 and C2.

2. Select **Stat** from the main menu.

3. Select **Nonparametrics**.

4. Select **Mann-Whitney**.

5. Make these entries in the dialog box:

  - Enter C1 for the first sample and C2 for the second sample.

  - Enter the confidence level in the indicated box. (A confidence level of 95.0 corresponds to a significance level of $\alpha = 0.05$.)

  - Select "alternate: not equal" in the indicated box; that selection refers to the alternative hypothesis, where "not equal" corresponds to a two-tailed hypothesis test.

Shown below are body mass index (BMI) measurements obtained from a random sample of men and women. (Ranks are shown in parentheses.) Also shown are the Minitab results that include a P-value of 0.3412, indicating that there is not a significant difference between the BMI measurements of men and women.

BMI Measurements

Men	Women
23.8 (11.5)	19.6 (2.5)
23.2 (9)	23.8 (11.5)
24.6 (14)	19.6 (2.5)
26.2 (17)	29.1 (22)
23.5 (10)	25.2 (15.5)
24.5 (13)	21.4 (5)
21.5 (6)	22.0 (7)
31.4 (24)	27.5 (19)
26.4 (18)	33.5 (25)
22.7 (8)	20.6 (4)
27.8 (20)	29.9 (23)
28.1 (21)	17.7 (1)
25.2 (15.5)	
$n_1 = 13$	$n_2 = 12$
$R_1 = 187$	$R_2 = 138$

```
Mann-Whitney Test and CI: Men, Women

               N   Median
Men           13   24.600
Women         12   22.900


Point estimate for ETA1-ETA2 is 1.900
95.3 Percent CI for ETA1-ETA2 is (-2.300,4.897)
W = 187.0
Test of ETA1 = ETA2 vs ETA1 not = ETA2
is significant at 0.3412
Test is significant at 0.3409 (adjusted for ties)
```

13-5 Kruskal-Wallis Test

The Triola textbook (excluding *Essentials of Statistics*) discusses the Kruskal-Wallis test and includes this definition.

Definition
The **Kruskal-Wallis test** (also called the *H* test) is a nonparametric test that uses ranks of sample data from three or more independent populations. It is used to test the null hypothesis that the independent samples come from populations with equal medians.

> H_0: The samples come from populations with equal medians.
> H_1: The samples come from populations with medians that are not all equal.

Caution: Using Minitab for a Kruskal-Wallis test is somewhat tricky, because we must enter all of the data in *one column*, then enter the corresponding identifiers in another column. For an example, see the sample data listed below and see the Minitab display showing the data listed in column C1 with identifiers in column C2. The Minitab display shows only the first ten data values, but all of the data must be listed with the same pattern.

If the data are listed in separate columns and you want to stack them as shown below, use **Data/Stack** (or **Manip/Stack** for Minitab Release 13 or earlier) to combine them into one big column. After selecting **Data/Stack**, you will get a "Stack Columns" dialog box. Enter the columns to be stacked, identify the column where the stacked data will be listed, and identify the column where the corresponding "subscripts" (identifiers) will be listed.

Weights (kg) of Poplar Trees

	\multicolumn{4}{c}{Treatment}			
	None	Fertilizer	Irrigation	Fertilizer and Irrigation
	0.15	1.34	0.23	2.03
	0.02	0.14	0.04	0.27
	0.16	0.02	0.34	0.92
	0.37	0.08	0.16	1.07
	0.22	0.08	0.05	2.38
n	5	5	5	5
$\bar{x}$	0.184	0.332	0.164	1.334
s	0.127	0.565	0.126	0.859

Stacked Data

↓	C1	C2-T
	Weight	Treatme
1	0.15	None
2	0.02	None
3	0.16	None
4	0.37	None
5	0.22	None
6	1.34	Fertilizer
7	0.14	Fertilizer
8	0.02	Fertilizer
9	0.08	Fertilizer
10	0.08	Fertilizer

After the sample data have been arranged in a single column with identifiers in another column, we can proceed to use Minitab as follows.

Procedure for the Kruskal-Wallis Program

1. Enter the data values for all categories in one column, and enter the corresponding identification numbers in another column.

2. Select **Stat** from the main menu.

3. Select **Nonparametrics**.

4. Select **Kruskal-Wallis**.

5. Make these entries in the dialog box:

- For Response, enter the column with all of the sample data combined.

- For Factor, enter the column of identification numbers or names.

- Click **OK**.

Here are the Minitab results obtained by using the Kruskal-Wallis test with the data in the preceding table:

Kruskal-Wallis Test on Weight

```
Treatment     N   Median  Ave Rank      Z
Fert&Irrig    5  1.07000      17.0   2.84
Fertilizer    5  0.08000       7.5  -1.31
Irrigation    5  0.16000       8.5  -0.87
None          5  0.16000       9.0  -0.65
Overall      20              10.5

H = 8.21  DF = 3  P = 0.042
H = 8.23  DF = 3  P = 0.041  (adjusted for ties)
```

Important elements of the Minitab display include the test statistic of $H = 8.21$ and the P–value of 0.042. We reject the null hypothesis (the samples come from populations with the same median) only if the P–value is small (such as 0.05 or less). Assuming a 0.05 significance level, we reject the null hypothesis of equal medians.

13-6 Rank Correlation

The Triola textbook introduces *rank correlation*, which uses ranks in a procedure for determining whether there is some relationship between two variables.

Definition
The **rank correlation test** (or Spearman's rank correlation test) is a nonparametric test that uses ranks of sample data consisting of matched pairs. It is used to test for an association between two variables, so the null and alternative hypotheses are as follows (where ρ_s denotes the rank correlation coefficient for the entire population):

H_0: $\rho_s = 0$ (There is *no* correlation between the two variables.)

H_1: $\rho_s \neq 0$ (There is a correlation between the two variables.)

First we describe the Minitab procedure, then we illustrate it with an example.

Minitab Procedure for Rank Correlation

1. Enter the paired data in columns C1 and C2.

2. If the data are already ranks, go directly to step 3. If the data are not already ranks, convert them to ranks by first using Minitab's **Rank** feature. (See Section 13-1 of this manual/workbook.) To convert the data to ranks, make these entries in the Rank dialog box:

   - Enter C1 in the box labeled Rank the data in.

   - Enter C1 in the box labeled Store ranks in.

   - Click **OK**.

   - Enter C2 in the box labeled Rank the data in.

   - Enter C2 in the box labeled Store ranks in.

   - Click **OK**.

3. Select **Stat** from the main menu.

4. Select **Basic Statistics**.

5. Select **Correlation**.

6. Enter C1 C2 in the dialog box, then click **OK**.

Consider the sample data in the table below. Because the paired data consist of ranks, the linear correlation coefficient r should not be used because it requires normal distributions, but the data consist of ranks that are not normally distributed. Instead, we use the rank correlation coefficient to test the claim that there is a relationship between the rankings of the two groups (that is, $\rho_s \neq 0$). Given below the table are the Minitab results.

Colleges Ranked by Students and *U.S. News and World Report*

College	Student Ranks	*U.S. News and World Report* Ranks
Harvard	1	1
Yale	2	2
Cal. Inst. of Tech.	3	5
M.I.T	4	4
Brown	5	7
Columbia	6	6
U. of Penn.	7	3
Notre Dame	8	8

Correlations: Student, U.S.News

```
Pearson correlation of Student and U.S.News = 0.714
P-Value = 0.047
```

The Minitab display shows that the test statistic is 0.714, and the P-value is 0.047. If we use a significance level of 0.05, the P-value of 0.047 is just slightly less than the significance level, so we *reject* the null hypothesis of no correlation, and we conclude that there is a correlation. (*Note:* If we use the critical value of $r_s = 0.738$ from Table A-9 in the textbook, the test statistic is less than the critical value, so we *fail to reject* the null hypothesis of no correlation, and we conclude that there is not a significant correlation. It is rare to have the conclusions disagree as they do here.)

13-7 Runs Test for Randomness

The Triola textbook (excluding *Essentials of Statistics*) discusses the runs test for randomness and includes these definitions.

Definitions
A **run** is a sequence of data having the same characteristic; the sequence is preceded and followed by data with a different characteristic or by no data at all.
The **runs test** uses the number of runs in a sequence of sample data to test for randomness in the order of the data.

When listing sequences of data to be used for the runs test, the textbook uses both numerical data and qualitative data, but Minitab works with numerical data only, so enter such a sequence using numbers instead of letters. As an illustration, consider the Boston rainfall for Mondays as listed in the Boston rainfall data set in Appendix B from the textbook. If we denote dry days (with 0.00 precipitation) by D's and rainy days (with precipitation amounts greater than 0.00) by R's, we get the following sequence of D's and R's, which we then convert to 0's and 1's as shown below.

```
DDDDRDRDDDRDDRDDDRDDRRRDDDDRDRDRRRDRDDDRDDDDRDRDDRDDDR

0000101001001000100111000010101110100010001010010001
```

Minitab Procedure for the Runs Test for Randomness

1. In column C1, enter the sequence of *numerical* data. (If the sequence of data consists of two qualitative characteristics, represent the sequence using 0's and 1's.)

2. Select **Stat** from the main menu.

3. Select **Nonparametrics**.

4. Select **Runs** Test.

5. Make these entries in the dialog box:

   • Enter C1 in the box for variables.

   • Click on the small circle for **Above and below**, and enter a number in the adjacent box. (Choose a number appropriate for the test. If using 0's and 1's, enter 0.5.)

   • Click **OK**.

Suppose that we want to use the runs test to determine whether there is sufficient evidence to support the claim that rain on Mondays is random. Using the preceding sequence of 0's and 1's, we get the following Minitab results. With a *P*-value of 0.140 we fail to reject the null hypothesis of randomness. Rain on Mondays in Boston appears to be random. Big surprise there.

```
Runs test for C1
Runs above and below K = 0.5
The observed number of runs = 30
The expected number of runs = 25.1154
19 observations above K, 33 below
P-value = 0.140
```

CHAPTER 13 EXPERIMENTS: Nonparametric Statistics

In Experiments 13–1 through 13–6, use Minitab's sign test program.

13-1. ***Testing for a Difference Between Reported and Measured Male Heights*** As part of the National Health and Nutrition Examination Survey conducted by the Department of Health and Human Services, self–reported heights and measured heights were obtained for males aged 12–16. Listed below are sample results. Is there sufficient evidence to support the claim that there is a difference between self–reported heights and measured heights of males aged 12–16? Use a 0.05 significance level.

Reported height	68	71	63	70	71	60	65	64	54	63	66	72
Measured height	67.9	69.9	64.9	68.3	70.3	60.6	64.5	67.0	55.6	74.2	65.0	70.8

P–value: _____

Conclusion: _____

_____

13–2. ***Testing for a Median Body Temperature of 98.6°F*** A pre–med student in a statistics class is required to do a class project. Intrigued by the body temperatures in Appendix B, she plans to collect her own sample data to test the claim that the mean body temperature is less than 98.6°F, as is commonly believed. Because of time constraints, she finds that she has time to collect data from only 12 people. After carefully planning a procedure for obtaining a simple random sample of 12 healthy adults, she measures their body temperatures and obtains the results listed below. Use a 0.05 significance level to test the claim that these body temperatures come from a population with a median that is less than 98.6°F.

97.6 97.5 98.6 98.2 98.0 99.0 98.5 98.1 98.4 97.9 97.9 97.7

P–value: _____

Conclusion: _____

_____

13–3. ***Testing for Median Underweight*** The Prince County Bottling Company supplies bottles of lemonade labeled 12 oz. When the Prince County Department of Weights and Measures tests a random sample of bottles, the amounts listed below are obtained. Using a 0.05 significance level, is there sufficient evidence to file a charge that the bottling company is cheating consumers by giving amounts with a median less than 12 oz?

11.4 11.8   11.7   11.0   11.9   11.9   11.5   12.0   12.1   11.9   10.9   11.3   11.5   11.5   11.6

P–value: _____

Conclusion: _____

_____

13–4. ***Nominal Data: Survey of Voters*** In a survey of 1002 people, 701 said that they voted in the recent presidential election (based on data from ICR Research Group). Is there sufficient evidence to support the claim that the majority of people say that they voted in the election?

P–value: _____

Conclusion: _____

_____

13–5. ***Nominal Data: Smoking and Nicotine Patches*** In one study of 71 smokers who tried to quit smoking with nicotine patch therapy, 41 were smoking one year after the treatment (based on data from "High-Dose Nicotine Patch Therapy," by Dale et al., *Journal of the American Medical Association,* Vol. 274, No. 17). Use a 0.05 significance level to test the claim that among smokers who try to quit with nicotine patch therapy, the majority are smoking a year after the treatment.

P–value: _____

Conclusion: _____

_____

13–6. ***Testing for Difference Between Forecast and Actual Temperatures*** Refer to the weather data set in Appendix B and use the actual high temperatures and the three day forecast high temperatures. (The Minitab workbook is WEATHER.) Does there appear to be a difference?

P–value: _____

Conclusion: _____

_____

13–7. ***Using Parametric Test*** Repeat Experiment 13–6 using an appropriate *parametric* test. (Use a *t* test for matched pairs.) Compare the results from the parametric test and the sign test. Do the results lead to the same conclusion? Is either test more sensitive to the differences between pairs of data?

_____

_____

13–8. ***Sign Test vs. Wilcoxon Signed–Ranks Test*** Repeat Experiment 13–6 by using the Wilcoxon signed-ranks test for matched pairs. Enter the Minitab results below, and compare them to the sign test results obtained in Experiment 13-6. Specifically, how do the results reflect the fact that the Wilcoxon signed-ranks test uses more information?

P–value: _____

Conclusion: _____

_____

Comparison: _____

_____

13–9. ***Testing for Drug Effectiveness*** Captopril is a drug designed to lower systolic blood pressure. When subjects were tested with this drug, their systolic blood pressure readings (in mm of mercury) were measured before and after the drug was taken, with the results given in the accompanying table (based on data from "Essential Hypertension: Effect of an Oral Inhibitor of Angiotensin-Converting Enzyme," by MacGregor et al., *British Medical Journal,* Vol. 2). Is there sufficient evidence to support the claim that the drug has an effect? Does Captopril appear to lower systolic blood pressure? Use the Wilcoxon signed–ranks test.

Subject	A	B	C	D	E	F	G	H	I	J	K	L
Before	200	174	198	170	179	182	193	209	185	155	169	210
After	191	170	177	167	159	151	176	183	159	145	146	177

P–value: _____

Conclusion: _____

_____

13–10. ***Are Severe Psychiatric Disorders Related to Biological Factors?*** One study used X-ray computed tomography (CT) to collect data on brain volumes for a group of patients with obsessive-compulsive disorders and a control group of healthy persons. The accompanying list shows sample results (in milliliters) for volumes of the right cordate (based on data from "Neuroanatomical Abnormalities in Obsessive-Compulsive Disorder Detected with Quantitative X-Ray Computed Tomography," by Luxenberg et al., *American Journal of Psychiatry,* Vol. 145, No. 9). Use the Wilcoxon rank–sum test with a 0.01 significance level to test the claim that obsessive-compulsive patients and healthy persons have the same median brain volumes. Based on this result, can we conclude that obsessive-compulsive disorders have a biological basis?

Obsessive-compulsive patients				Control group			
0.308	0.210	0.304	0.344	0.519	0.476	0.413	0.429
0.407	0.455	0.287	0.288	0.501	0.402	0.349	0.594
0.463	0.334	0.340	0.305	0.334	0.483	0.460	0.445

P–value: _____

Conclusion: _____

_____

13–11. ***Testing the Anchoring Effect*** Randomly selected statistics students were given five seconds to estimate the value of a product of numbers with the results given in the accompanying table. Is there sufficient evidence to support the claim that the two samples come from populations with different medians? Use the Wilcoxon rank–sum test.

Estimates from Students Given $1 \times 2 \times 3 \times 4 \times 5 \times 6 \times 7 \times 8$

1560	169	5635	25	842	40,320	5000	500	1110	10,000
200	1252	4000	2040	175	856	42,200	49,654	560	800

Estimates from Students Given $8 \times 7 \times 6 \times 5 \times 4 \times 3 \times 2 \times 1$

100,000	2000	42,000		1500	52,836	2050	428	372	300	225
64,582	23,410	500	1200	400	49,000	4000	1876	3600	354	750
640										

P–value: _____

Conclusion: _____

_____

13-12. ***Does Weight of a Car Affect Head Injuries in a Crash?*** Data were obtained from car crash experiments conducted by the National Transportation Safety Administration. New cars were purchased and crashed into a fixed barrier at 35 mi/h, and measurements were recorded for the dummy in the driver's seat. Use the sample data listed below to test for differences in median head injury measurements (in hic) among the four weight categories. Is there sufficient evidence to conclude that head injury measurements for the four car weight categories are not all the same? Do the data suggest that heavier cars are safer in a crash?

Subcompact:	681	428	917	898	420
Compact:	643	655	442	514	525
Midsize:	469	727	525	454	259
Full–Size:	384	656	602	687	360

P–value: _____

Conclusion: _____

_____

13-13. ***Do All Colors of M&Ms Weigh the Same?*** Refer to the M&M data set in Appendix B from the textbook. At the 0.05 significance level, test the claim that the weights of M&Ms have the same median for each of the six different color populations. If it is the intent of Mars, Inc., to make the candies so that the different color populations are the same, do your results suggest that the company has a problem that requires corrective action?

P–value: _____

Conclusion: _____

_____

13-14. ***Homerun Distances*** Refer to the "Homeruns" data set in Appendix B from the textbook. Consider the homerun distances to be samples randomly selected from populations. Use a 0.05 significance level to test the claim that the populations of distances of homeruns hit by Barry Bonds, Mark McGwire, and Sammy Sosa have the same median.

P–value: _____

Conclusion: _____

_____

13–15. ***Business School Rankings***   *Business Week* magazine ranked business schools two different ways. Corporate rankings were based on surveys of corporate recruiters, and graduate rankings were based on surveys of MBA graduates. The table below is based on the results for 10 schools. Is there a correlation between the corporate rankings and the graduate rankings? Use a significance level of $\alpha = 0.05$.

School	PA	NW	Chi	Sfd	Hvd	MI	IN	Clb	UCLA	MIT
Corporate ranking	1	2	4	5	3	6	8	7	10	9
Graduate ranking	3	5	4	1	10	7	6	8	2	9

P–value: _____

Conclusion: _____

_____

13–16. ***Correlation Between Restaurant Bills and Tips*** Students of the author collected sample data consisting of amounts of restaurant bills and the corresponding tip amounts. The data are listed below. Use rank correlation to determine whether there is a correlation between the amount of the bill and the amount of the tip.

Bill (dollars)	33.46	50.68	87.92	98.84	63.60	107.34
Tip (dollars)	5.50	5.00	8.08	17.00	12.00	16.00

P–value: _____

Conclusion: _____

_____

13–17. ***Correlation Between Heights and Weights of Supermodels*** Listed below are heights (in inches) and weights (in pounds) for supermodels Niki Taylor, Nadia Avermann, Claudia Schiffer, Elle MacPherson, Christy Turlington, Bridget Hall, Kate Moss, Valerie Mazza, and Kristy Hume. Is there a correlation?

Height	71	70.5	71	72	70	70	66.5	70	71
Weight	125	119	128	128	119	127	105	123	115

P–value: _____

Conclusion: _____

_____

13-18. ***Buying a TV Audience*** The *New York Post* published the annual salaries (in millions) and the number of viewers (in millions), with results given below for Oprah Winfrey, David Letterman, Jay Leno, Kelsey Grammer, Barbara Walters, Dan Rather, James Gandolfini, and Susan Lucci, respectively. Is there a correlation?

Salary	100	14	14	35.2	12	7	5	1
Viewers	7	4.4	5.9	1.6	10.4	9.6	8.9	4.2

P–value: _____

Conclusion: _____

_____

13-19. ***Cholesterol and Body Mass Index*** Refer to Data Set 1 in Appendix B from the textbook and use the cholesterol levels and body mass index values of the 40 women. Is there a correlation between cholesterol level and body mass index?

P–value: _____

Conclusion: _____

_____

13-20. ***Testing for Randomness of Survey Respondents*** When selecting subjects to be surveyed about the Blue Fang *Zoo Tycoon* game, the subjects were selected in a sequence with the genders listed below. Does it appear that the subjects were randomly selected according to gender?

M M F F F M F M M M M M F F M M F F F F M F

P–value: _____

Conclusion: _____

13–21. ***Testing for Randomness in Dating Prospects*** Fred has had difficulty getting dates with women, so he is abandoning his strategy of careful selection and replacing it with a desperate strategy of random selection. In pursuing dates with randomly selected women, Fred finds that some of them are unavailable because they are married. Fred, who has an abundance of time for such activities, records and analyzes his observations. Given the results listed below (where M denotes married and S denotes single), what should Fred conclude about the randomness of the women he selects?

M M M M S S S S S S M M M M M M S S S M M M M M M M M M S S S

P–value: _____

Conclusion: _____

_____

13–22. ***Testing for Randomness of Baseball World Series Victories*** Test the claim that the sequence of World Series wins by American League and National League teams is random. Given below are recent results with American and National league teams represented by A and N, respectively. What does the result suggest about the abilities of the two leagues?

N A A A N N A A N N N A A A N A

N A N A A A N A N A A A N A N A

P–value: _____

Conclusion: _____

_____

13–23. ***Testing for Randomness of Presidential Election Winners*** For a recent sequence of presidential elections, the political party of the winner is indicated by D for Democrat and R for Republican. Does it appear that we elect Democrat and Republican candidates in a sequence that is random?

R R D R D R R R D D R R R D D D D D R R D D R R D R R R D D R

P–value: _____

Conclusion: _____

_____

13–24. ***Testing for Randomness of Baseball World Series Victories*** Test the claim that the sequence of World Series wins by American League and National League teams is random. Given below are recent results, with American and National League teams represented by A and N, respectively.

A N A N N N A A A A N A A A A N A N N A A N N A A A A A N A N

N A A A A A N A N A N A N A A A A A A A N N A N A N N A A N

N N A N A N A N A A A N N A A N N N N A A A N A N A N A N A A A

N A N A A A N A N A

P–value: _____

Conclusion: _____

_____

13–25. ***Testing for Randomness of Odd and Even Digits in Pi*** A *New York Times* article about the calculation of decimal places of π noted that "mathematicians are pretty sure that the digits of π are indistinguishable from any random sequence." Given below are the first 100 decimal places of π. Test for randomness of odd (O) and even (E) digits. Test for randomness of odd and even digits.

1 4 1 5 9 2 6 5 3 5 8 9 7 9 3 2 3 8 4 6 2 6 4 3 3 8 3 2 7 9 5 0 2 8 8 4 1 9 7 1

6 9 3 9 9 3 7 5 1 0 5 8 2 0 9 7 4 9 4 4 5 9 2 3 0 7 8 1 6 4 0 6 2 8 6 2 0 8 9 9

8 6 2 8 0 3 4 8 2 5 3 4 2 1 1 7 0 6 7 9

P–value: _____

Conclusion: _____

_____

14

Statistical Process Control

14-1 Run Charts

In the Statistical Process Control chapter in the Triola textbook (excluding *Essentials of Statistics*) we define **process data** to be data arranged according to some time sequence, such as the data in the table below. At the Altigauge Manufacturing Company, four altimeters are randomly selected from production on each of 20 consecutive business days, and the table lists the *errors* (in feet) when they are tested in a pressure chamber that simulates an altitude of 1000 ft. The error measurements are displayed in the table with bold text. On day 1, for example, the actual altitude readings for the four selected altimeters are 1002 ft, 992 ft, 1005 ft, and 1011 ft, so the corresponding errors (in feet) are 2, −8, 5, and 11. Those four errors have a mean of 2.50 ft, a median of 3.5 ft, a range of 19 ft, and a standard deviation of 7.94 ft.

Aircraft Altimeter Errors (in feet)

Day					Mean	Median	Range	St. Dev.
1	2	-8	5	11	2.50	3.5	19	7.94
2	-5	2	6	8	2.75	4.0	13	5.74
3	6	7	-1	-8	1.00	2.5	15	6.98
4	-5	5	-5	6	0.25	0.0	11	6.08
5	9	3	-2	-2	2.00	0.5	11	5.23
6	16	-10	-1	-8	-0.75	-4.5	26	11.81
7	13	-8	-7	2	0.00	-2.5	21	9.76
8	-5	-4	2	8	0.25	-1.0	13	6.02
9	7	13	-2	-13	1.25	2.5	26	11.32
10	15	7	19	1	10.50	11.0	18	8.06
11	12	12	10	9	10.75	11.0	3	1.50
12	11	9	11	20	12.75	11.0	11	4.92
13	18	15	23	28	21.00	20.5	13	5.72
14	6	32	4	10	13.00	8.0	28	12.91
15	16	-13	-9	19	3.25	3.5	32	16.58
16	8	17	0	13	9.50	10.5	17	7.33
17	13	3	6	13	8.75	9.5	10	5.06
18	38	-5	-5	5	8.25	0.0	43	20.39
19	18	12	25	-6	12.25	15.0	31	13.28
20	-27	23	7	36	9.75	15.0	63	27.22

A *run chart*, which is a sequential plot of *individual* data values over time, can be generated as follows.

1. Enter all of the data (in sequence) in column C1.

2. Select the main menu item of **Stat**.

3. Select the subdirectory item of **Quality Tools**.

4. Select the option of **Run Chart**.

5. Make these entries in the dialog box:

- Select **Single column** and enter C1 in the adjacent box.

- Enter 1 for the subgroup size (because we want *individual* values plotted).

- Click **OK**.

Using the process data in the preceding table, the above procedure will result in this run chart:

Minitab

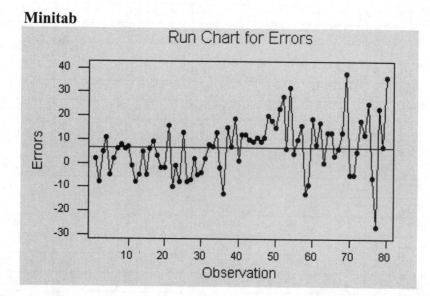

Examine the above run chart and note that it reveals this problem: As time progresses from left to right, the heights of the points appear to show a pattern of increasing variation. See how the points at the left fluctuate considerably less than the points farther to the right. The Federal Aviation Administration regulations require errors less than 20 ft (or between 20 ft and −20 ft), so the altimeters represented by the points at the left are OK, whereas several of the points farther to the right correspond to altimeters not meeting the required specifications. It appears that the manufacturing process started out well, but deteriorated as time passed. If left alone, this manufacturing process will cause the company to go out of business. That is, the run chart suggests that the process is not **statistically stable** (or **within statistical control**) because it has a pattern of increasing variation.

14-2 *R* Charts

The textbook describes *R* charts as sequential plots of ranges. Using the data from the preceding table, for example, the *R* chart is a plot of the ranges 19, 13, 15, . . . , 63. *R* charts are used to monitor the *variation* of a process. The procedure for obtaining an *R* chart is as follows.

1. Enter all of the data in sequential order in column C1.

2. Select **Stat** from the main menu.

3. Select **Variable Charts for** Subgroups. (If using Minitab Release 13 or earlier, select the subdirectory item of **Control Charts**.)

4. Select the option of **R**.

5. Make these entries in the dialog box:

   - Enter C1 in the box for the column containing all observations.

   - Enter a subgroup size of 4 in the indicated box. (We use 4 here because the Table 14-1 data have 4 observations each day.)

   - Click the **R options** button, click on the Estimate tab, and click on the **Rbar** button. (If using Minitab Release 13 or earlier, verify that the **R bar estimate** is being used in the calculation of the upper and lower control limits. Click on **Estimate**, then click on the small circle labeled **Rbar**. Click **OK**.)

Using the data in the preceding table, the Minitab display will be as shown below.

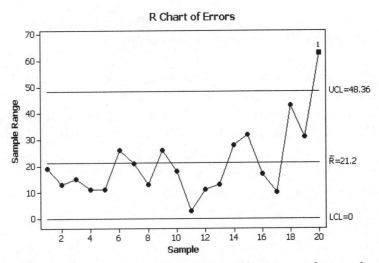

We can interpret the above *R* chart by applying these three out-of-control criteria given in the textbook:

1. There is no pattern, trend, or cycle that is obviously not random.

2. No point lies beyond the upper or lower control limits.

3. There are not 8 consecutive points all above or all below the center line.

Analyzing the above R chart leads to the conclusion that the process *is out of statistical control* because the second criterion is violated: There is a point beyond the upper control limit. Also, there is a pattern of an upward trend.

14-3 $\overline{x}$ Charts

We can now proceed to use the same altimeter error data listed in the table from Section 14-1 of this manual/workbook to create an $\overline{x}$ chart. The textbook explains that an $\overline{x}$ chart is used to monitor the *mean* of the process. It is obtained by plotting the sample means. Follow these steps to generate an $\overline{x}$ chart.

1. Enter all of the data in sequential order in column C1.

2. Select **Stat** from the main menu.

3. Select **Variable Charts for** Subgroups. (If using Minitab Release 13 or earlier, select the subdirectory item of **Control Charts**.)

4. Select the option of **Xbar**.

5. Make these entries in the dialog box:

   - Enter C1 in the box for the column containing all observations.

   - Enter a subgroup size of 4 in the indicated box. (We use 4 here because the data in the table consist of 4 observations each day.)

   - Click **Xbar options**, click the **Estimate** tab, and click on the **Rbar** button. (If using Minitab Release 13 or earlier, verify that the **R bar estimate** is being used in the calculation of the upper and lower control limits. Click on **Estimate**, then click on the small circle labeled **Rbar**. Click **OK**.)

Using the data in table of altimeter errors, the Minitab display will be as shown below.

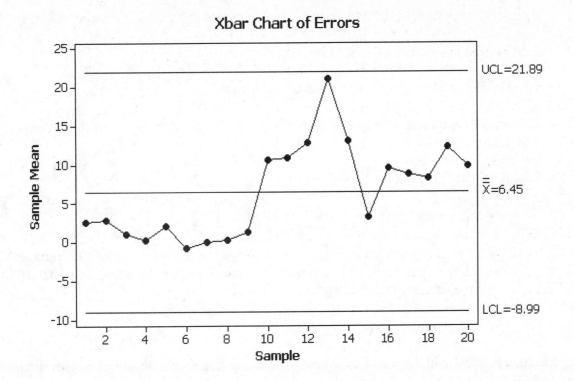

Xbar Chart of Errors

We can interpret the $\bar{x}$ chart by applying the three out-of-control criteria given in the textbook. We conclude that the mean in this process is *out of statistical control* because there does appear to be a pattern of an upward trend, and the last of the following conditions is also violated:

1. There is no pattern, trend, or cycle that is obviously not random.

2. No point lies beyond the upper or lower control limits.

3. There are not 8 consecutive points all above or all below the centerline.

14-4 *p* Charts

A control chart for attributes (or *p* chart) can also be constructed by using the same procedure for *R* charts and $\bar{x}$ charts. A *p* chart is very useful in monitoring some process proportion, such as the proportions of defects over time. As an example, the Altigauge Manufacturing Company produces altimeters in batches of 100, and each altimeter is tested and determined to be acceptable or defective. Listed below are the numbers of defective altimeters in successive batches of 100.

Defects: 2 0 1 3 1 2 2 4 3 5 3 7

Follow these steps to generate a Minitab *p* chart:

1. Enter the numbers of defects (or items with any particular attribute) in column C1.

2. Select the main menu item of **Stat**.

3. Select the subdirectory item of **Control Charts**.

4. Select the option of **Attribute** Charts. (If using Minitab Release 13 or earlier, select **P** (for p chart).

5. Make these entries in the dialog box:

    • Enter C1 in the box labeled Variable.

    • Enter the subgroup size. (For example, if C1 consists of the numbers of defects per batch and each batch consists of 100 altimeters, then enter 100 for the subgroup size.)

    • Click **OK**.

The Minitab display will be as shown below. We can interpret the control chart for p by considering the three out-of-control criteria listed in the textbook. Using those criteria, we conclude that this process is out of statistical control for this reason: There appears to be an upward trend. The company should take immediate action to correct the increasing proportion of defects.

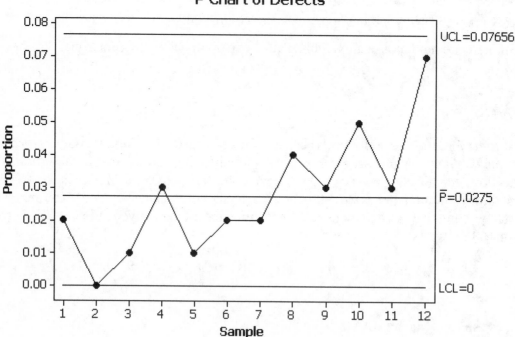

CHAPTER 14 EXPERIMENTS: Statistical Process Control

Constructing Control Charts for Aluminum Cans *Experiments 1 and 2 are based on the axial loads(in pounds) of aluminum cans that are 0.0109 in. thick, as listed in the "Axial Loads of Aluminum Cans" data set in Appendix B of the textbook. An axial load of a can is the maximum weight supported by its side, and it is important to have an axial load high enough so that the can isn't crushed when the top lid is pressed into place. The data are from a real manufacturing process, and they were provided by a student who used an earlier version of this book.*

14–1. **R Chart** On each day of production, seven aluminum cans with thickness 0.0109 in. were randomly selected and the axial loads were measured. The ranges for the different days are listed below, but they can also be found from the values given in Appendix B of the textbook. (The Minitab worksheet is *CANS*.) Construct an *R* chart and determine whether the process variation is within statistical control. If it is not, identify which of the three out-of-control criteria lead to rejection of statistically stable variation.

78 77 31 50 33 38 84 21 38 77 26 78 78
17 83 66 72 79 61 74 64 51 26 41 31

14–2. **x̄ Chart** On each day of production, seven aluminum cans with thickness 0.0109 in. were randomly selected and the axial loads were measured. The means for the different days are listed below, but they can also be found from the values given in Appendix B of the textbook. (The Minitab worksheet is *CANS*.) Construct an x̄ chart and determine whether the process mean is within statistical control. If it is not, identify which of the three out-of-control criteria lead to rejection of statistically stable variation.

252.7 247.9 270.3 267.0 281.6 269.9 257.7 272.9 273.7 259.1
275.6 262.4 256.0 277.6 264.3 260.1 254.7 278.1 259.7 269.4
266.6 270.9 281.0 271.4 277.3

14-3. **Weights of Minted Quarters** The U.S. Mint has a goal of making quarters with a weight of 5.670 g, but any weight between 5.443 g and 5.897 g is considered acceptable. A new minting machine is placed into service and the weights are recorded for quarters randomly selected every 12 min for 20 consecutive hours. The results are listed in the following table. Use MINITAB to construct a run chart. Determine whether the process appears to be within statistical control.

Hour	Weight of Quarter (grams)					Mean	Range
1	5.639	5.636	5.679	5.637	5.691	5.6564	0.055
2	5.655	5.641	5.626	5.668	5.679	5.6538	0.053
3	5.682	5.704	5.725	5.661	5.721	5.6986	0.064
4	5.675	5.648	5.622	5.669	5.585	5.6398	0.090
5	5.690	5.636	5.715	5.694	5.709	5.6888	0.079
6	5.641	5.571	5.600	5.665	5.676	5.6306	0.105
7	5.503	5.601	5.706	5.624	5.620	5.6108	0.203
8	5.669	5.589	5.606	5.685	5.556	5.6210	0.129
9	5.668	5.749	5.762	5.778	5.672	5.7258	0.110
10	5.693	5.690	5.666	5.563	5.668	5.6560	0.130
11	5.449	5.464	5.732	5.619	5.673	5.5874	0.283
12	5.763	5.704	5.656	5.778	5.703	5.7208	0.122
13	5.679	5.810	5.608	5.635	5.577	5.6618	0.233
14	5.389	5.916	5.985	5.580	5.935	5.7610	0.596
15	5.747	6.188	5.615	5.622	5.510	5.7364	0.678
16	5.768	5.153	5.528	5.700	6.131	5.6560	0.978
17	5.688	5.481	6.058	5.940	5.059	5.6452	0.999
18	6.065	6.282	6.097	5.948	5.624	6.0032	0.658
19	5.463	5.876	5.905	5.801	5.847	5.7784	0.442
20	5.682	5.475	6.144	6.260	6.760	6.0642	1.285

14-4. ***Minting Quarters: Constructing an R Chart***   Using the same process data from Experiment 14-3, construct an R chart and determine whether the process variation is within statistical control. If it is not, identify which of the three out-of-control criteria lead to rejection of statistically stable variation.

14-5. ***Minting Quarters: Constructing an $\bar{x}$ Chart***   Using the same process data from Experiment 14-3, construct an $\bar{x}$ chart and determine whether the process mean is within statistical control. If it is not, identify which of the three out-of-control criteria lead to rejection of a statistically stable mean. Does this process need corrective action?

14-6. ***p Chart for Deaths from Infectious Diseases***   In each of 13 consecutive and recent years, 100,000 children aged 0–4 years were randomly selected and the number who died from infectious diseases is recorded, with the results given below (based on data from "Trends in Infectious Diseases Mortality in the United States," by Pinner et al., *Journal of the American Medical Association,* Vol. 275, No. 3). Use MINITAB to construct a p chart. Do the results suggest a problem that should be corrected?

Number who died: 30 29 29 27 23 25 25 23 24 25 25 24 23

14-7. ***p Chart for Victims of Crime***   For each of 20 consecutive and recent years, 1000 adults were randomly selected and surveyed. Each value below is the number who were victims of violent crime (based on data from the U.S. Department of Justice, Bureau of Justice Statistics). Use MINITAB to construct a p chart. Do the data suggest a problem that should be corrected?

29 33 24 29 27 33 36 22 25 24 31 31 27 23 30 35 26 31 32 24

14-8. ***p Chart for Boston Rainfall***   Refer to the Boston rainfall amounts in the data set from Appendix B of the textbook. Delete the last value for Wednesday, so that there are 52 weeks of seven days each. For each of the 52 weeks, let the sample proportion be the proportion of days that it rained. In the first week for example, the sample proportion is $3/7 = 0.429$. Use MINITAB to construct a p chart. Do the data represent a statistically stable process?

14-9. ***p Chart for Marriage Rates***   Use p charts to compare the statistical stability of the marriage rates of Japan and the United States. In each year, 10,000 people in each country were randomly selected, and the numbers of marriages are given for eight consecutive and recent years (based on United Nations data).

Japan: 58 60 61 64 63 63 64 63

United States: 98 94 92 90 91 89 88 87

Minitab Release 14
Important Menu Items

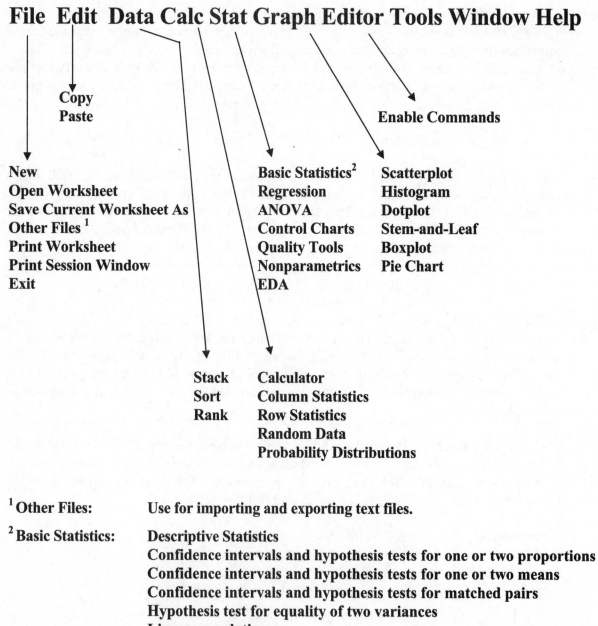

File Edit Data Calc Stat Graph Editor Tools Window Help

Copy
Paste

Enable Commands

New   Basic Statistics[2]   Scatterplot
Open Worksheet   Regression   Histogram
Save Current Worksheet As   ANOVA   Dotplot
Other Files [1]   Control Charts   Stem-and-Leaf
Print Worksheet   Quality Tools   Boxplot
Print Session Window   Nonparametrics   Pie Chart
Exit   EDA

Stack   Calculator
Sort   Column Statistics
Rank   Row Statistics
          Random Data
          Probability Distributions

[1] Other Files: Use for importing and exporting text files.

[2] Basic Statistics: Descriptive Statistics
                        Confidence intervals and hypothesis tests for one or two proportions
                        Confidence intervals and hypothesis tests for one or two means
                        Confidence intervals and hypothesis tests for matched pairs
                        Hypothesis test for equality of two variances
                        Linear correlation
                        Test for normality

Retrieving Data from the CD-ROM

The CD-ROM packaged with the Triola textbook has worksheets for the data in Appendix B of the textbook. Refer to Appendix B in the textbook for the worksheet names. Here is the procedure for retrieving a worksheet from the CD-ROM:

1.   Click on Minitab's main menu item of **File**.

2.   Click on the subdirectory item of **Open Worksheet**.

3.   You should now see a window like the one shown below.

4.   In the "Look in" box at the top, select the location of the stored worksheets. For example, if the CD-ROM is in drive D and you want to open the Minitab worksheet COLA.MTW, do this:

   • In the "Look in" box, select drive **D** (or whatever drive contains the CD-ROM).
   • Double click on the folder **DataSets**.
   • Double click on the folder **Minitab**.
   • Click on **COLA.MTW** (or any other worksheet that you want).

5.   Click on the **Open** bar.

6.   You will get a message that a copy of the contents of the file will be added to the current project. Click **OK**. The columns of data should now appear in the Minitab display, and they are now available for use.

Index